Praise for *The Hungry Gh*

"More than any baker I know, Jonathan embraces the unpredictability of bread-making. Indeed, as revealed in the pages of this book, he makes these challenges into a kind of poetic partnership. The dirty little secret is that we professional bakers are in constant dance and strife with uncooperative grains, reluctant ovens, and either stilted or exuberant fermentation. Our *victory* is that we manage, by hook or by crook, to have things looking mostly right when the bakery door is unlocked. This beautiful book will serve as both guidance and reassurance to the aspiring baker that we are all aspiring bakers."

— BARAK OLINS, James Beard Award–winning baker, ZU Bakery

"Through story, poems, and recipes, Jonathan Stevens reminds us why we bake and offers us the tools to become better bakers along the way: MORE HYDRATION, MORE FERMENTATION, MORE HEAT IN THE OVEN! Beautifully written, with profound ruminations on baking as an act of (Neolithic) revolution. Brilliant!"

— JENNIFER LAPIDUS, founder, Carolina Ground; author of *Southern Ground*

"Bread is a fun, vast topic whose technical side can intimidate. In *The Hungry Ghost Bread Book*, Jonathan Stevens finds so much pleasure in bread and makes baking so wholly approachable that you might never realize how grounded the recipes are in the ways that a professional makes bread really delicious."

— EDWARD BEHR, editor, *The Art of Eating*

"This is a beautifully written book by a true artisan. Jonathan Stevens has been refining his techniques and recipes for decades at Hungry Ghost Bread in Northampton, Massachusetts. He opens the book with a riveting account of the rhythms and textures of the bakery's daily grind, then breaks down the different breads with clear and concise recipes for making them at home and even suggestions of alternative uses for overproofed dough. Easy to read and likely to inspire, this book will take your bread-making to the next level."

— SANDOR ELLIX KATZ, fermentation revivalist; author of *The Art of Fermentation* and other fermentation bestsellers

"It's impossible to read through the recipes in *The Hungry Ghost Bread Book* without being inspired to scoop out some sourdough starter and get mixing. And while you may open this book initially for the recipes—of which there are plenty of great ones—you'll leave with a mind full of sound processes, techniques, and a new set of baker's sensibilities."

— MAURIZIO LEO, author of James Beard Award–winning *The Perfect Loaf*

"*The Hungry Ghost Bread Book* is a throwback to the classics that inspired me to become a baker. This book is far more than its many excellent recipes. It's about caring about your ingredients, caring about the craft, and most of all, making the good stuff and feeding the people. I am inspired to bake all over again."
— MARK DYCK, host, Rise Up! The Baker Podcast; founder, Bakers4Bakers

"Reader, this book is one of a kind. The author knows his stuff, but he's not just teaching—he's doing. He's wrapping his arm around you and slipping you into his bread-woven life. I truly think he's something of a witch (the fermentation he conjures definitely has the aura of enchantment). Go put a handful of unbleached flour into a bowl and wet it with a good splash of well water—and see who comes calling…"
— JOHN THORNE, author of James Beard Award–winning *Pot on the Fire*

"Artisan bread pioneers Cheryl Maffie and Jonathan Stevens have been indoctrinating bakers and bread lovers into the Church of Gluten for over twenty years. *The Hungry Ghost Bread Book* offers a unique look into the philosophy and methods behind their beloved breads. Theirs has long been one of my favorite bakeries anywhere, and now the joys of Hungry Ghost are accessible far beyond the Pioneer Valley."
— ANDREW JANJIGIAN, head baker, Wordloaf; author of *Breaducation*

"You might not be able to get to Jonathan's world-class Hungry Ghost Bread bakery, but you can make these recipes, which are simple enough for beginners yet will inspire even die-hard bread-heads. Just reading this marvelous cookbook is half the fun, because Jonathan leavens his recipes with decades of hard-earned knowledge, from sourdough science to eccentric lore, puns, and poetry: 'even a bad day baking / beats merely making dough.' Amen!"
— ANDREW COE, award-winning food historian; bread fanatic

"Jonathan Stevens leads you into his head and his work, escorting you with stories and poems through Hungry Ghost Bread's great recipes. As you learn how to make good naturally leavened bread—and what to do when the dough goes awry—you'll learn how this lively baker and bakery came to be, too."
— AMY HALLORAN, regional grain advocate; author of *The New Bread Basket*

"I struggled to write a blurb for this book—and not because I didn't like it. It's just that every time I tried to write a few words, I discovered they were already written by Jonathan. I kept coming back to the three 'mores' of bread baking: more hydration, more fermentation, more heat! This book is enjoyable and inspiring to read, with attention not only to the bread, but to family and community, too."
— THOMAS LEONARD, author of *The Bread Book*

THE
HUNGRY GHOST
BREAD BOOK

THE
HUNGRY GHOST
BREAD BOOK

*An Offbeat Bakery's Guide to
Crafting Sourdough Loaves, Flatbreads,
Crackers, Scones, and More*

JONATHAN STEVENS
Foreword by RICHARD MISCOVICH

Chelsea Green Publishing
White River Junction, VT
London, UK

First published by Chelsea Green Publishing | PO Box 4529 | White River Junction, VT 05001 | West Wing, Somerset House, Strand | London, WC2R 1LA, UK | www.chelseagreen.com

A Division of Rizzoli International Publications, Inc. | 49 West 27th Street | New York, NY 10001 | www.rizzoliusa.com

Gruppo Mondadori | Via privata Mondadori, 1 | 20054 Segrate (Milano), ITA

All illustrations copyright © 2024 by Maya Malachowski Bajak.

Lyrics on page 86 are from "Corn Meal Dance," words by William Parker, copyright © 2007 by Centering Music (BMI).

Developmental Editor: Natalie Wallace
Copy Editor: Karen Wise
Proofreader: Nancy A. Crompton
Indexer: Linda Hallinger
Designer: Melissa Jacobson

ISBN 9781645022602 (paperback) | ISBN 9781645022619 (ebook)
Library of Congress Control Number: LCCN 2024021168 (print) | LCCN 2024021169 (ebook)

Our Commitment to Green Publishing

Chelsea Green sees publishing as a tool for cultural change and ecological stewardship. We strive to align our book manufacturing practices with our editorial mission and to reduce the impact of our business enterprise in the environment. We print our books using vegetable-based inks whenever possible. This book may cost slightly more because it was printed on paper that contains recycled fiber, and we hope you'll agree that it's worth it. *The Hungry Ghost Bread Book* was printed on paper supplied by Marquis that is made of recycled materials and other controlled sources.

Instagram: @ChelseaGreenBooks and @RizzoliBooks | Facebook: @ChelseaGreenPub and @RizzoliNewYork | X: @ChelseaGreen and @Rizzoli_Books | YouTube: @ChelseaGreenPub and @RizzoliNY

Printed in Canada.
First printing August 2024.
10 9 8 7 6 5 4 3 2 1   24 25 26 27 28

Contents

Foreword

Jonathan Stevens is inside a bubble of fermentation friends and colleagues I've known for so long, I can't exactly remember when and where we met. I know *about* when—after Y2K but before September 11. It might have been at an oven-building workshop led by Alan Scott at Noah Elbers's Orchard Hill Breadworks, but it doesn't really matter. That first meeting of like-minded bakers making naturally leavened, whole-grain, wood-fired bread led to lifelong spontaneous fermentation, just like the bubbling containers of sourdough starter you have in your kitchen . . . or will, once you finish Jonathan's inspiring book.

Our lives have crisscrossed like a challah, braided with various strands: hangouts at several (often snowbound) cabins in the Northeast, at sunny summer events with a bunch of other enthusiastic bakers, and, most notably, at Hungry Ghost Bread itself. My first visit to Hungry Ghost was on April 10, 2004, when my Johnson & Wales University colleague Ciril Hitz and I drove from Providence, Rhode Island, to Northampton, Massachusetts, to help celebrate our friends' *ouverture officielle de la boulangerie*. When we arrived at the grand opening, the place was packed—a milieu of eager friends and locals ready to embrace this temple as a community hub. It wasn't because there was free bread, even though there was. It wasn't because of the puppets and the kids making art, though I recall those included in the festivities, too. It was because Jonathan and Cheryl had already spent years connecting with and distributing goodness to their community.

I became aware that Hungry Ghost was dedicated to feeding not only the appetites but also the souls, brains, and hearts of the symbiotic, mycorrhizal network that *continues* to circle back to downtown Northampton. Hungry Ghost has helped that network grow. Jonathan was one of the first bakers I knew who sought out local farmers and millers, a pioneer and benefactor of the now-vibrant local grain economy.

In the ensuing decades, the bread at Hungry Ghost has continued day after day, year after year, into the oven, endlessly baking. Jonathan's intimate proximity to the act of baking, often while the beloved community slumbers, has led to his personal, textured, and well-articulated description of how to make bread. As a culinary educator and bread-book author, I know it isn't easy to describe how to shape a loaf, but get a load of this: ". . . our approach is to handle the dough without actually touching it too much. Or at least, never too long in any one spot." Those words are a primary rule of dough handling, no matter how many folds, tucks, revolutions, flips, and sound effects you use to shape a loaf. There are many ways to shape a boule, but all successful approaches include not touching it too long in any one place.

Here are Jonathan's words regarding one of his tenets for increasing dough hydration: "Pinch the dough to get a sense of its slackness or tightness. Add more water if needed so it's not too sloppy, but not too stiff. Steer into the wind a bit, but don't let the sail luff. Added water is like the wind, and you are trimming the dough toward a distant point on the horizon. . . ." When I first read that, I felt as though I was gazing through a new door of bakeception. I couldn't decide if I should allow myself to get choked up or simply stand up and go make some bread. (I did both.) Explore the flexible formulas, flirt with the parameters of the process, and, in the end, enjoy your food, whether it is a loaf of bread, a scone, a cracker, or a savory meal based on the starches of your satisfying labor.

There is a powerful message to anybody reading this book, illuminated by the second word of the book's subtitle: *offbeat*. It is true, Hungry Ghost *is* offbeat, in exactly the way Jonathan, Cheryl, and Family have conceived of it and brought it to life. How many bakeries include a daily baker-penned poem with each loaf?

To some extent, all baking is offbeat in the way that the baker tailors it to their life and influences the outcome. Even though we work as servants to grain, yeast, enzymes, temperature, and the unstoppable swoosh of time's wingèd chariot, we still influence the process. Once you understand how sourdough bread is rooted in a beautiful system of interrelated variables, you'll be able to establish your own beat to produce bread in your own kitchen, with your own tools, in a cranky *or* friendly oven, and (hopefully) with grain grown in the region where you live.

Lest you miss it, this is a punny book. Copious cultural references and Jonathan's baking playlist illuminate the rich life experiences that have made him who he is. By the time you finish the book, you'll probably have been inspired to create your own diverse production playlist. In that way, this book is definitely right on the beat.

Through the art, craft, and science of naturally leavened baking, Jonathan and Family have created Hungry Ghost Bread and helped pioneer the community bakery, which brings local grain and yeast together in an alchemical metamorphosis to educate, inspire, and compel appreciative, loyal, and knowledgeable bread lovers. I feel fortunate to be swinging around the sun during the American Bread Renaissance, along with a hard-working, homespun, appropriately detail-oriented and irreverent band of sisters and brothers. Hungry Ghost isn't the only community bakery to arise in the evolved food-scape of America in the last thirty years, but it is the only one that does things in the way it does. This is your chance to experience a bit of it yourself.

Lines written on Roosevelt St., Rhode Island
Early Summer 2024
Richard Miscovich

Acknowledgments

No one bakes or eats a loaf all on their own. For over twenty years, I have been baking in tandem with Cheryl Maffie: my partner, confidante, coparent, co*grand*parent, business collaborator, and lover. She coaxed our bakery into being two decades ago and is the one who has kept it going all these years. I just bake the bread, but she actually *runs the business!* This book is a reflection of our work together and would not have been remotely possible without her input, ingenuity, acumen, and faith in things unseen.

Our kids—Ana, Nia, Elijah, and Haden—have all played a tremendous part, from scraping paint off the floor before we first opened to mixing dough and preparing tax statements. They have put up with us again and again and again . . . and I love them for it.

Our beautiful and suffering coworkers have often had to face the reality of being part of a family-owned business. I ask for their forgiveness and their blessing.

We have relied on many a skilled friend to help us fix a mixer or repair part of the oven or replace a broken compressor. The bread would not get baked without these people: Rich Heiman, Dennis Lombardi, Chris Strom, Eric Huther, Stephen Yoshen, Tyler Stosz, David Holmes, and many, many others.

Home-baking tests for these formulas were generously (and methodically) carried out by Avery Hart of Skokie, Illinois, a retired doctor and new friend who gave me crucial feedback on both the recipe details and the manuscript in general.

At 10 a.m., when the curtain goes up in Hungry Ghost, we remember why we're here: there are customers to serve. To all the regulars who introduced themselves and asked how we were doing. To the ones who shared a book or some music or a piece of good news. To the ones who brought their kids or their grandkids or their friends and made us part of their day: that's the reciprocity that I was hoping for.

Introduction

even a bad day
baking / beats
merely making dough

Every time a baker refreshes their starter, every time flour and water mix to make a dough, every time a round is shaped and watched over and carried to the mouth of the hearth, every time an oven gives off the aroma of baking bread and draws hungry people to its warmth, the Neolithic Revolution is renewed. For over ten thousand years we humans have been grinding grain, making slurries, and coaxing them closer to the heat so there will be something good to eat together. Bread is the plate, the napkin, and the bulk of almost any traditional meal.

This ritual, this reenactment, is not merely symbolic. The elements are real, the nourishment is physical—the war against hunger is not theater. The building blocks of true civilization, flour and clean running water and village hearths, are part of our inheritance, but they must be respected and put to constant use. Renewed. Rituals become empty when their meanings are forgotten, their movements become mere gestures, their medicines too weak for our newfangled dis-eases.

Bread is ancient medicine, indeed, and one that still cures all kinds of ailments: hunger, blues, loneliness, and alienation, to name a few. Just try eating a whole loaf all by yourself! As many have pointed out, the very word *companion* comes from the Latin root *panis*, or bread. So, "company" was who you broke bread with. That is why the Neolithic Revolution still needs to be defended to this day: it represents an oasis between the nomadic tribalism of the past and the technofeudalism of the seemingly inescapable future. If we are to offer our friends, our families, and our neighbors (of any persuasion) a gift of substance, fresh bread will be always be a great place to start.

And where *do* we start? Perhaps with a nod to wheat: the grass that our ancestors in the Fertile Crescent slowly domesticated over thousands of plant generations, selecting the biggest, juiciest seeds, the husks that would be free-threshing but not shatter. That certainly took effort and forethought. The wheat we take for granted as an agricultural product is analogous to the cats and dogs we share our homes with.

Wheat is so ubiquitous we hardly notice it, like oxygen, but it's just as essential and, unfortunately, just as easily squandered and polluted. Much of the wheat ingested these days is hidden in processed foods in the form of "vital gluten." This is certainly one of the biggest culprits in engendering gluten intolerance. Personally, I love gluten because I couldn't make bread without it. The T-shirt I'm wearing right now says POWERED BY GLUTEN! Our bakery is a church of gluten. And yet, our bread is digestible only because it is also *naturally leavened*. Gluten without fermentation is not digestible! Wheat naturally contains phytic acids, which impair nutrient absorption, but fermentation (inoculation with wild yeast and bacteria) actually neutralizes this food inhibitor. Yet another reason why the old revolution (Neolithic) has better principles than the new one (be it the Industrial or the Technological). This is absolutely crucial to understand: sourdough bread is not merely a "style" of bread. It *is* bread. That other stuff—yeasted bread and "straight" dough—that stuff may look like bread, and to some it may even taste like bread, but it is not truly digestible.

Aztec people have long understood that for corn to be truly digestible it must be nixtamalized, or chemically treated with limestone or ash. This process not only deactivates mycotoxins, but it also unbinds the grain's niacin, thus helping to prevent the disease of pellagra. In our analogous coevolution with corn's sister grain, wheat, it is the wheat slurry's anchoring of wild yeasts and beneficial bacteria that makes all the difference. If you leave a mix of flour and water out for a week, it will start to bubble! A combination of what's in the air and in the wheat and in your hands and mouth will take root. It is perhaps in our role as "wild yeast farmers" (as one agriculturalist friend puts it) that bakers really make their mark.

It is somehow easier or more romantic for those of us who are descendants of Europeans to recognize sanctity in *other* people's grain traditions. Many First Nations all over the Americas still do corn dances and sacred plantings, profound acknowledgments of their gratitude for the sustenance of maize. Rice, too, has its rituals, one of which is to free Hungry Ghosts from their particular realm of Hell. That is the one our bakery is named for, because one day about forty years ago, in the motley Buddhist sangha I still belong to, Kato-Shonin and Clare-Anjusan laid out an extra plate at our communal table. It was for the Hungry Ghost, an evil twin of Passover's Elijah who is invited to eat but is definitely not benevolent. It is the spirit of insatiable desire, the ghost of those who, like all of us at times, can never be satisfied. The Hungry Ghosts live in an afterlife, with big stomachs and tiny little mouths. They are horrible to behold. And yet, for me, they beg the questions: Who eats when I eat? And who does not? What am I feeding when I stuff my mouth? And can we ever free the suffering spirits of both the living and the dead with a single, well-intentioned loaf?

Well, probably not, and while Kato-Shonin continues to enjoy our bread (especially the French Bâtard!), he insists that the name of the shop is too scary. What I take as coincidental affirmation of our name is that on the day we first opened, April 10, 2004, there was a lecture on Hungry Ghosts by a Taiwanese scholar at the Smith College Museum of Art a mere block away. I rushed up between bakes to learn what I could and to make an offering.

My personal apprenticeship in the ritual of bread-making happened in between these two points: building the New England Peace Pagoda in Leverett, Massachusetts, with the Nipponzan Myohoji monastic order in my twenties and building a bakery twenty years later, across the valley, in Northampton. "Working on a building," as the old gospel song says. I backed my way into it, really, doing everything the hard way and getting plenty of bumps, scrapes, burns, and bruises. After trying my hand at hanging drywall, practicing social work, and songwriting, the only boss I knew how to work for was a little baby; as a stay-at-home dad (twice over), I learned the useful focus of early-morning baking.

With little Elijah, and then Haden, bouncing on the countertop at 5 a.m., I would have a planned project for us to do together. *The Tassajara Bread Book*'s Country Loaf or a lemon-cranberry muffin. When the boys got big enough to go to school, I built a cob oven outside the house and started scaling up the recipes. Then we built an Alan Scott–style brick oven in our walk-out basement and I really got to work. That first big day, after driving all the way to Portland, Maine, to buy a used mixer from a deconsecrated church, I thought I was ready.

I had never used a big mixer before. Flour and water and starter were poured in. The orbital arm began spinning, and I moved in to drip a little more water from the pitcher in my hand. . . . Well, the pitcher was glass (big mistake!) and the hook grabbed it and seconds later there were sharp shards all throughout the dough. Death Bread! Big World Premier! I pressed the OFF button and went back to bed.

Part of what this book is all about is to help you avoid some of these indignities. Sure, the school of hard knocks might build character, but it rarely makes a decent loaf of bread before the first few years are out. The Neolithic Revolution is also about the beginnings of written culture and passing down information that can help invent *new* wheels, instead of the same old ones, over and over. In my case, the dough and the oven did their work on me, and I survived.

Most of the books I found (with the exception of Alan Scott and Daniel Wing's *The Bread Builders*) were not that helpful. Most enlightening were a couple of retreats that Alan hosted out at the Headlands Center for the

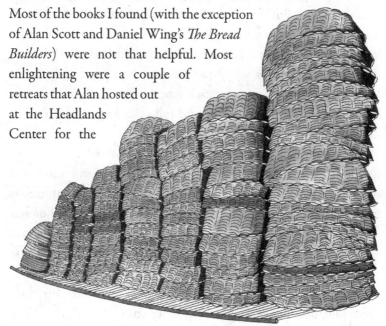

Arts in Marin, California, where a bunch of us start-up artisans could compare notes and get some real peer education. RIP, Alan Scott!

So I baked along, starting a little wholesale business out of my home. I fed the starter, mixed the dough, proofed it in 5-gallon containers, and baked it in a beautiful but fussy 3 × 4-foot wood-fired black brick oven (black ovens are fired inside the baking chamber). I had to get up halfway through the night to rake the embers forward; in spite of this, hearth temps would never quite even out. By midday, I delivered the still-warm loaves to various community-supported farms, co-ops, restaurants, and markets. The CSAs were the best— the most welcoming and financially worthwhile outlets, where I could sometimes graze for strawberries. The restaurants were fine with accepting dented or less-attractive loaves (they were going to get cut up anyway). But the grocery stores really did me in: with little to no incentive to sell my bread, they would often be holding leftover loaves, which waited to shame me whenever I delivered. All I could do was take them home and throw them on the compost pile. As the CSAs wound down for winter each year, I swore that I would somehow open a retail shop and get paid full price for each piece while never having to drive around delivering again.

For a couple of years, I integrated my business with a community nonprofit in Holyoke, expanded my wholesale distribution, and refined the recipes. Then Cheryl and I found a storefront to rent in Northampton and started scrounging for that other kind of dough. We sold "bread futures" to customers who already knew us (a kind of early crowdfunding); we found a sympathetic banker; we got a loan from Cheryl's mom.

Why is this important? Because context is everything: bread is a social phenomenon as much as a culinary one. Before the Neolithic

Revolution, food was shared strictly within a tight kin structure around the campfire or in a tent. Since the Dot-Com Revolution, most food is gobbled from take-out containers in front of glowing screens. In our version of events, friends and strangers alike are invited to take part in the economy of baking. After twenty years in the same spot, it still astounds me: I have become a village baker, watching customers' kids grow up and their parents age. We *know* people or get to meet them, and bread is the medium.

All four of our kids have worked at the shop at one time or another. Two of them still do! Nia's kids are in school now, and they drop by to run around outside or dance in the flour dust on the floor or stick their just-washed hands into the gushy dough. Many of our former coworkers are still in touch; some are close friends whose kids also come by to grab a free cookie. This is what a real economy looks like! Sometimes it includes cash or credit cards, but more importantly, it often includes an exchange of greetings, news, or opinions; a comment on the music playing; or simply a nod of the head or a raised eyebrow. We don't make much money, but bread is the currency we "print," and it has value. Sometimes the exchange value of a given thing is not realized until much later. And I don't mean with interest. I mean in human relationships, which are, after all, our only real assets.

Even in home baking, there is an economy, though it is often the gift economy. You give away or share your loaves with friends, family, and neighbors. Perhaps you are lucky enough to buy grain from a farmer you know, or flour from a local mill. These are invaluable connections. Scaling them up can certainly be tricky, but not necessarily corrosive. Our bakery is open seven days a week (that's what keeps the wood-fired oven warm, the starter fed, and the flywheel spinning), so of course we operate thanks to a whole crew of employees. Theirs is tough work, full of nuances and picayune instructions from the likes of me. We try to pay them as much as we can while still keeping the business afloat; that's what Cheryl does. It is a challenging time to be a wage worker with the high cost of living, and we don't take them for granted. Too much of the mythology of bread baking has an individualistic bent to it, a pioneering "I did it all myself" sort of image. But even if you *could* farm the land yourself, harvest the wheat alone, and then mill it and somehow find time to bake it off, you would still need other people to buy it and eat it, let alone build the tractor you used or sell you the seed or play the music you listened to all along the way.

Our grain economy happily includes a local grain mill, just down the road in Holyoke, called Ground Up Grain. Andrea and Christian Stanley source regional grain and use the hydropower from South Hadley Falls to run a beautiful pair of New American Stone Mills built by Andrew Heyn in Elmore, Vermont. Their driver, Blue, delivers flour to us every Friday. That's part of our grain economy. So's my friend Matt Littrell, at Hillcrest Foods, a professional baker who knows about what other kinds of flour we need. So are my *compatri-otes, la famille Beauchemin*, who founded and still run La Milanaise organic flour company in southern Quebec. They single-handedly reinvigorated wheat growing in eastern Canada and now export a fabulous, affordable product.

Another part of our grain economy emerged spontaneously out of the Covid pandemic. We suddenly had to pivot (like everyone else) and minimize close contact with our customers while maintaining some semblance of sales and cash flow. We closed the front door and opened a take-out window. Instead of taking orders ahead of time (something we've never been good at!), we took the radical but self-evident route to just ask customers to pay us online . . . *after they got home!* As a financial risk, one $6 loaf or so at a time, it seemed like a no-brainer. Cumulatively, it was a bigger leap, but we went for it and it really paid off. Turns out that most people, after an initial shock, appreciate being trusted! Whatever little losses we incurred through forgetfulness or lack of computer skills were more than compensated for by reflected goodwill and some generous donations. This experiment went on successfully for a year and a half—another utopian theorem proved correct!

So here's my exchange with you, dear fellow baker: for the meager admission price to a book version of our bakery, I'm offering up all the information that I can. The numbers, the sequence, the pacing, and the feel of the process as well as I can put it into words. Why? Because personally, I can bake only so many loaves a day anyway. Because, despite a constitutional resistance to learning things the conventional way, I think I've stumbled onto a technique that makes excellent bread, and I want to share it. We'll cover an array of differ-ent flours, mixing techniques, how to shape slack dough, and how to come up with a "dough management" style that will deliver the best bread without sacrificing the baker's whole day. In the first part of the book, we'll cover general strategy and approaches to this style of bread-making. In the second part of the book, we'll delineate precise formulas. There are sixteen formulas for "standard" bread loaves here; five different flatbreads; four of our folds; five types of crackers; and some scones, biscuits, bagels, and recipes to use up stale bread.

Allow me to state it clearly and boldly: there are three things to do when making this style of bread, and each starts with "more." MORE HYDRATION, MORE FERMENTATION, MORE HEAT IN THE OVEN! That's it, in a nutshell: my philosophy of bread.

Almost all the bread I come across, in bakeries and in books, is under-hydrated, underfermented, and underbaked. Why? Because it's easier to handle, because blasting steam at an underproofed loaf makes it "pop" in a sexy way, and because people are afraid of pushing the envelope. Don't be afraid: the worst than can happen (and it will!) is a mess of dough. If you want to see the view, you've got to stand closer to the edge. Excellence happens within inches of that edge.

Baking bread is certainly anachronistic in the twenty-first century, but so is riding a bike (well, at least a non-e-bike) or playing an acoustic guitar. I do it because it seems useful and it engages my whole body: lifting 50-pound sacks of flour, pouring 50-pound bins of water, wrestling huge blobs of dough, scoring loaves at the oven's mouth, diving long wooden peels onto the steaming hearth, then splitting wood for the fire or sweeping the floor. Four days a week, the bakery contains my restlessness.

As good as it might be, a loaf can be appreciated only in its obliteration. It is made to be eaten, "translated" into eater. It does a baker good to be rid of their loaves. The lack of accumulation is part of the flow: after mixing's done, we discard the surplus starter. It is both valuable and superfluous. The ritual of bread-making, daily or weekly, is one of practice and alchemy, not of procuring actual gold. The transmutation occurs in both the ingredients and the practitioner, if you allow it.

For two thousand years, ancient Greeks held a festival known as the Eleusinian Mysteries. Eleusis, just 14 miles from Athens (and now, apparently, an industrial zone), was home to the Rharian Field, where Triptolemus grew the first plot of wheat after the goddess Demeter gifted him the seed. Upwards of 3,000 people would stream out of the city, dip into the ocean, fast for days, and then cross the Bridge of Jests.

A psychoactive brew of barley and mint (and possibly ergot) would keep them dancing all night, down into the underground theater of the Telesterion, where women would carry the first fruits of the harvest and make the *pelanos*, or sacred bread. Emerging finally into daylight, mirroring Demeter's rescue of Persephone, the initiates would be transformed: each now a seed themselves, threshed through ceremony, unafraid of Hades and his realm of death. Spores of leaven, every soul.

A Day in the Life
of a Loaf

*counting the hours
in hearth-spins
and weeks in oven-loads*

Any loaf, or any baker for that matter, is more than the sum of its parts: I call it gestalt baking. Your attention is constantly refocusing from the details to the whole and back again, from the fineness of the flour to an array of baked loaves and the line of customers waiting to claim them. The dots are joined, but never in a straight line—rather like a pointillist painting of a working bakery. Here's how our dots are joined, through a wide fish-eye lens.

Because the bakery is open seven days a week, because the wood-fired oven is never off, and because everything is on a flywheel, our production schedule has a certain Möbius-strip quality: when I get in at 6:30 a.m., the oven is hot, the starter is bubbling, and the dough (shaped, in baskets) is in the walk-in, ready to be baked. Or so we hope. . . .

There's always a competing list of things to do first: wash hands, check the oven temp, open up the starter bucket and peer inside, glance at the retarded dough in the cooler and judge its volume. Turn on some music! Kill a fly and wash hands again. Open some windows, wipe down the glass door on the oven.

Oven temp? 518°F (270°C). Perfect. Pour four or five pitcherfuls of water into the steam chamber and get it set. Roll out the speed racks from the walk-in, wipe down the insides of the rack's zip-down plastic covers, and get a good look. Does the basketted dough seem ripe, like a piece of fruit ready to pluck, but not going to fall of its own accord? Plump, but not overinflated: that's what we're hoping for, and if we judged the overnight trajectory well, that's where we are. If not, a little "floor time" (time to sit on the counter). If it's overfermented, score it lightly or not at all. If it deflates and dies on the peel, toss it into the compost.

Though this 2.5-meter Llopis oven with the rotating hearth has perfectly even heat throughout, whatever goes in first is blasted with high temps, and we have to work quickly and efficiently while the dial descends. Whole wheat Eight-Grain and Rye loaves to begin with, followed by French Bâtards, and chased finally by the Rosemary Bread. Fill in the gaps, but don't let the loaves touch (or "kiss," in the French terminology). Dip the razor blade in hot water, score the loaves with their individual branding.

Sweaty work, but fun. About sixty-five loaves fit in at a time.

While the first bake is in (but after the rest of the dough is returned to the walk-in and the floor is swept), *then* we start mixing dough for the next day. Scaling out starter (which should be actively bubbling and viscous), checking water temps, grabbing 50-pound bags of flour. Different days have different mixes, but the ratios of water and salt and starter are fairly constant. The flours and added herbs or grains switch up: rosemary, beer-soaked rye flakes, malted barley, shredded beets.

Each dough gets two mixes. The first mix is for 4 minutes on first speed, to incorporate everything, except the salt. Then a 20-minute rest (or *autolyze*, as the French say), during which the flour is passively absorbing water, and then a second mix on second speed, also for 4 minutes. This is when the salt gets added, and more water, carefully. In the last minute of mixing, herbs or other inclusions are spun into that cyclone of dough. Then the colossus (our largest mixes weigh in at 120 pounds) is removed from the mixer and put in rectangular bins, which are set aside. The next mix is prepped, and so on.

Don't forget the bread in the oven! Turn on the exhaust fan and let some steam out of the room. The first loaves will start to come out soon after that. Optimal baking time per loaf? Forty minutes. But let the bread tell you when it's ready: tap it on the bottom and listen for that hollow sound. Slide it onto a cooling rack.

On the way back to the mixer, take a sip of coffee. Turn up the music, switch it from opera to jazz or something else. There are anywhere from four to eight mixes a day to do.

Once the first bake is done and the hearth is swept out, it's time to start another fire to heat up the oven for the next batch. It'll take about an hour and a half. Open the damper, push back the ashes from the clean-out vent, open the firebox, and get the kindling going with a match to an empty flour bag. Build up the fire with chunkier splits of wood until it's roaring. Go outside and scope out the chimney—make sure you're not smoking up the neighborhood!

Now that there's a minute, throw some of yesterday's bread in the toaster and find some butter and jam. Sit and eat it, without looking at a screen. Then back to the mixer and the firebox.

So this goes on for a while. Coworkers dance around each other and clean out baskets and scrape couches (proofing cloths) and bring hot pastries from the back. At 10 a.m., someone hangs the YES! sign outside and customers file in. Some are regulars, like Tony, aka

"Happy Scone Man." Some are newbies and curious about the place: *Do you have gluten-free bread? Why is it all sourdough? What is a French Bastard?*

The oven is heating up, the loaves are disappearing off the cooling racks. The phone is ringing. Dough in the bins is given an "envelope" fold three to four times throughout the day. Once the oven is back to temp—500°F (260°C), at least—then another round of steam is set and the loaves are delivered to their fate. This cycle happens anywhere from three to five times a day. Up to 400 loaves on our busiest days (not counting Thanksgiving!).

Around 3 p.m., there's usually a tandem bake while we start to shape the next day's dough. In this case, *tandem* means that there's no firing of the oven. The residual heat will be sufficient, either because there isn't much to bake or the type of dough requires less heat, such as an Olive-Semolina Fougasse (white flour, lots of surface area) or Challah (enriched dough with eggs and olive oil). Steam is less necessary, too.

The team has mobilized, the bench (big table) is cleared, and the small scale is ready. The binned doughs have expanded voluminously, pressing up against their lids. Time to shape it! The big blob is hoisted up, folded one more time, and rolled over, ready to be cut to size. Each loaf is weighed to 1¾ pounds (794 g) so that when it loses a little water during the bake, it'll still be the standard 1½ pounds (680 g). Five, sometimes six of us are jammed around the bench, brandishing our dough knives, grabbing blobs of dough, and swirling them into rounds. Dusting copiously with flour (it's slack dough with a long time before the next morning) to allow for expansion. Seam-side down. Baskets six to a sheet pan in the plastic-jacketed speed racks.

We talk, we make silly puns (or I do), we hustle and do the next dough and the next. Some doughs get special dusters, such as oats or corn grits or quinoa flakes. The French dough is shaped into bâtards (literally, a cross between a baguette and a boule), nested into their couches, and dusted with a mix of white and rice flour.

Some dough, if running a little warm, will go into the cooler with the plastic rack cover flap open to cool down a bit. Others roll right in, zipped down and ready for their long, dreamy nap. Occasionally, a dough will get a little more floor time to ensure proper development. It's all educated guesswork, with some leeway for mistakes and the rare disaster. Underproofed dough can be salvaged. Overproofed, not so much.

After shaping and cleaning up, most of us go home. The closer stays behind, does some wood-chopping, fires up the oven for the next day, and, most crucially, feeds the starter right before going home so that it'll be ready in the morning. The flywheel has been successfully spun one more time, the momentum sustained, the bread cart rolling on to tomorrow.

Tools of the Trade

Demeter
she uses
de metric system

That glass water pitcher I used for my first attempt at bread was definitely a mistake, and while I'm not much of a gearhead with baking or biking or even guitar playing, I have found that avoiding specifically designed equipment can equally be a mistake.

To make bakery-quality sourdough bread at home is not an expensive undertaking, but it does require certain tools. The following list is divided into two categories: "Essential" and "Helpful but Not Essential." You probably already have some of these items in your kitchen.

Essential

DIGITAL KITCHEN SCALE. One of the keys to success in bread making is to weigh the ingredients in metric units. Durable, reliable digital kitchen scales are inexpensive.

INSTANT-READ DIGITAL THERMOMETER. A digital thermometer with a stainless steel probe is useful for checking the temperatures of your water and dough. Seasoned bakers can work without thermometers, but those with less experience should incorporate temperature checks into their baking routine.

MIXING BOWL. A capacity of at least 5 quarts (4.73 L) is a good size for the recipes in this book. If you mix by hand, you can use a mixing bowl made out of almost anything—metal, ceramic, glass, or plastic; in some parts of the world, home bakers mix dough in a wooden trough. If you use a stand mixer, you'll use the bowl that comes with it.

FERMENTATION CONTAINER. The dough will slowly rise during several hours of bulk fermentation, the first round of proofing between mixing and shaping. At Hungry Ghost, we ferment the dough in very large rectangular plastic bins. A polycarbonate food pan with a flat lid makes an excellent scaled-down home version. This is a standard item at restaurant supply stores. The straight sides and transparent material allow you to accurately monitor how much the dough has risen, and the rectangular shape is convenient for folding the dough. For the recipes in this book, the ideal configuration is a half-size pan—roughly 12 × 10 inches (30.5 × 25.4 cm) and 4 inches (10 cm) deep.

Alternatively, any covered bowl or tub with a capacity of about 5 quarts (4.73 L) is a time-honored fermentation container. The cover is needed to keep the dough from drying out. The main disadvantage of using a bowl rather than a straight-sided clear container is that it's harder to be precise in gauging how much the dough has risen.

A WARM SPOT. For bulk fermentation, you'll need a warm spot to park the dough while it ferments. The ideal temperature is 78 to 80°F (25.5 to 26.5°C). You may have a good spot in your home, though it can vary from season to season. The oven light provides an inexpensive solution. This light, typically a 40-watt incandescent bulb, will warm the oven slightly. If the oven light makes it too warm, then bring the temperature down a bit with either of these hacks: (1) prop the oven door slightly ajar with a pencil (or a piece of dowel) or (2) place a small pot of ice water in the oven and replenish the ice as needed. And if the oven light doesn't warm the oven enough, bring the temperature up a bit by placing a small pot of just-boiled water inside; replenish the hot water as needed.

Alternatively, a seedling heat mat or some other flat, gentle heat pad can be placed on the counter underneath your fermentation container. This solution keeps your oven available for food preparation while your dough ferments.

PLASTIC DOUGH SCRAPER. This tool is invaluable for scraping dough out of a mixing bowl or a fermentation bin.

STAINLESS STEEL DOUGH KNIFE. The dough knife is a rectangular piece of stainless steel with a handle, used to divide and manipulate the dough after bulk fermentation. You could use a kitchen knife to divide the dough, but for other tasks you'll still need a dough knife.

PROOFING BASKET WITH CLOTH LINER. After the dough has been divided and shaped into loaves, each loaf will be placed into a proofing basket (or *banneton*), where it will complete the fermentation process before baking. Proofing baskets are typically made from coiled or woven rattan. They are typically sold with a removable liner of linen or

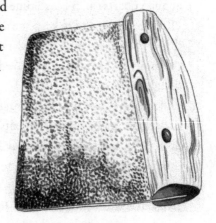

cotton. For most of the bread formulas in this book, which yield two loaves with a dough weight of around 1¾ pounds (794 g) each, the best match will be round baskets with a diameter of 8 or 9 inches (20–23 cm). You can also improvise a proofing basket by lining a plastic basket or colander with a tea towel.

RAZOR BLADE. A single-edge razor blade is useful for scoring the bread just before it goes into the oven. You can also use a baker's lame, a double-edge razor blade attached to a small handle.

STEAM SETUP. At professional bakeries like Hungry Ghost, the oven is set up to inject steam at the start of a bake. By bathing the dough in steam, we delay formation of the crust, which allows the bread to rise more while baking (this is known as oven spring). There are two methods to improvise a steam setup at home: (1) enclose the dough in a baking vessel or (2) generate steam by boiling water in a steam pan near the dough.

A baking vessel traps and holds water vapor that escapes from the dough, enveloping the bread in steam as it bakes. A cast-iron Dutch oven is a good choice for a baking vessel. You can use any Dutch oven that you may already have in your kitchen, as long as it's made entirely of metal, with or without an enamel coating. If you're looking to buy a Dutch oven, the design best suited for bread-making is a combo cooker, so called because the lid doubles as a skillet. For the recipes in this book, a combo cooker with capacity of about 3 liters (3.2 qt) will hold the dough nicely, with enough headroom for the dough to expand while baking. Note that a round baking vessel can be used only to bake round loaves, which is the shape of most breads in this book.

Alternatively, a steam pan fills the whole oven with steam. For this method you just need an all-metal skillet, such as cast iron.

Fill it with water, put it on the lowest rack of the oven, and let
it preheat along with the oven. A steam pan can be used when
baking breads of any shape.

HEAT-RESISTANT OVEN MITTS. Wear good silicone oven mitts.
Don't try to handle cast-iron Dutch ovens at 450°F (232°C) using
cloth mitts, pot holders, or folded kitchen towels.

WIRE COOLING RACK. When your bread comes out of the oven,
it should cool with air circulating freely around the entire loaf.

STARTER JARS. Get some straight-sided ("shoulderless"), wide-mouth
glass jars with lids to hold your starter. The straight-sided design
makes it easier to take out the starter. It's useful to have different
sizes, a couple of pint jars and a couple of half-pint jars. You can buy
new mason jars or simply reuse empty jam jars or canning jars.

Helpful but Not Essential

SILICONE BAKING MAT. Any smooth, clean countertop in your
kitchen will be adequate for bread-making. At Hungry Ghost, we
work at a big butcher-block baker's table. Butcher block is an ideal
surface because it holds a fine dusting of flour in place, to reduce
sticking of dough. You can get the same result at home by placing
a silicone pastry mat on your countertop. A rectangular pastry
mat that covers about 4 square feet—for example, 28 × 20 inches
(71 × 51 cm)—is a good size.

BAKING STONE. A rectangular ceramic baking stone is useful
for several purposes. It provides an even distribution of heat for
hearth-style baking directly on the stone. It also buffers the drop
in temperature when you open the oven door by providing thermal
ballast within the oven. Choose a size that allows 2 inches (5 cm) or
more of free space between the stone and the oven wall on each side.

OVEN PEEL. If you bake on a baking stone or other flat surface, rather than in a baking vessel, then you'll need something to transfer the dough into the oven. An oven peel, either wooden or metal, is the baker's tool for this purpose. A rimmed baking sheet, turned upside down, can double as an improvised peel.

DIGITAL THERMOMETER WITH AIR PROBE. A digital thermometer with an air probe will allow you to continuously monitor the temperature inside your oven, whether you're using it for baking or for bulk fermentation. Most ovens do not match the settings on their dials, and this two-piece gizmo (probe inside, display out) will give a far more accurate reading.

STAND MIXER. At Hungry Ghost we mix batches of dough weighing up to 100 pounds (45 kg), and we use a big commercial floor mixer to do it. For home baking, a stand mixer is entirely optional. Many home bakers mix sourdough entirely by hand, and some small-scale professional bakers do as well. If you already own a stand mixer with a bowl capacity of at least 5 quarts (4.73 L), it should be able to handle all the recipes in this book. Smaller units may not have the needed power or capacity.

PASTRY BRUSH. Just the tool for brushing off loaves on their way into the oven (and removing excess flour from cracker dough). On Fridays, we use a pastry brush to paint an egg-wash on the Challah.

For making crackers, the following items are helpful but not essential:

ROLLING PIN. If you make crackers or fresh pasta, you'll need something for rolling them out. I prefer an old-fashioned, American-style heavy-duty rolling pin with a center rod and handles on each end.

DOUGH DOCKER. This utensil is a spiked roller mounted on a handle. It is used to prick the dough to prevent blistering and minimize rising of crackers and some flatbreads.

PASTRY WHEEL. A pastry wheel will either lightly score or cut right through your cracker dough, depending on the degree of pressure used.

OFFSET SPATULA. This tool is handy for picking up thinly rolled cracker dough and transferring it to a peel.

Getting Started with "Ghost Farts"

this big bucket
belches back
good morning!

When Bea brings a group of kids from the People's Institute day care to visit, they all want to know, "Where's the Ghost?"

I tell them, "It's right here, in this bucket!"—the white, 5-gallon sourdough home. Then I take off the lid and give them a whiff and show them how it's bubbling. "That's the Ghost, farting! This *whole bakery* runs on Ghost farts!"

And it's true. Bread is a fermented food, and as sourdough bread bakers, we're essentially wild yeast farmers, keeping a population of microbes well nourished and inoculating each new batch of dough with a bunch of tiny critters. Ghost farts rule.

A starter (aka leaven, biga, poolish, chef, Mother) is a symbiotic community of wild yeasts and Lactobacilli maintained in a medium of wheat flour and water. The wild yeasts, such as *Saccharomyces cerevisiae* (brewer's yeast), are single-cell fungi that are all around us. The bacteria, such as the famously named *Lactobacillus sanfranciscensis*, are more adapted to specific environments. These are the predigesters of the dough, the carbon dioxide expellers, the tiny magicians that ferment our mixes and prepare them to be baked into actual food.

How old is any given leaven? I'm not sure it really matters much: it's not the pedigree or whether your grandma used it, but how you're treating it and what you're feeding it that counts. It's never older than the last time you added water and flour. The "magic" is not in its human history, but in what's floating in the air and carried on the flour dust. We are giving those yeast cultures and those bacteria a happy place to be, inside our buckets. We are collaborating with them, and they have domesticated us, just like the wheat itself, or apple trees or horses.

I took part in Rob Dunn's Global Sourdough Project out of the University of North Carolina. We sent off a sample of starter to be analyzed, along with dozens and dozens of other bakers from around the country. I didn't expect anything surprising, but when I looked at the data map online, I noticed that our starter was oddly different from other bakeries nearby, even ones that I knew used the very same flour. The dominant yeast in our sample was *Kazachstania humilis*, which only three other sites on his national data map had. I've never even been to Kazakhstan!

Mysteries and magic, via science. Why do we leaven the dough with this difficult, challenging, demanding stuff? Because it gives us healthy bread, since only natural fermentation neutralizes the phytic acids that occur in wheat, rendering it more digestible to humans. This is something that is all too rarely discussed in the nutritional literature. As with any fermented food, some of the "digestion" has happened before we even put it in our mouths. Cows can eat grass out of the field because they have four stomachs (and they still fart a lot). As bipedal mammals, we gained a lot of mobility, got these wacky big brains, language, opposable thumbs, and . . . we have to ferment and cook our grains.

Quick yeasts, Fleischmann's, and everything else that comes with industrial baking are a big part of the reason we have so much wheat intolerance. "Straight dough," or bread leavened with anything besides what we call sourdough, can smell like and taste sort of like bread, but it will never be digested in the same way. In fact, I would contend that our naturally fermented white loaves are healthier than a straight-dough whole wheat, any day of the week. My mantra, if you haven't heard already, is that sourdough bread is *not* a "style" of bread: it *is* real bread, and has been for over 10,000 years. Good bread depends on a lot of things, including the wheat farmer, the miller, the oven-builder, and the baker. But without the leaven, it would not be the cornerstone of our sustenance.

At Hungry Ghost, we always feed our starter with a one-quarter portion of rye flour (for tang) and the remaining three-quarters is white flour. We happily give away small containers of our starter to customers, knowing that somehow it would be wrong to charge for mere "essence." We keep our starter buckets out of the cooler, except during summer vacation. I say "Good morning" to it every day, and it always bubbles back at me.

We feed the Ghost twice a day: first, a snack, or maintenance feed, after depleting the bucket when mixing's done (in the early afternoon) and again last thing at night with a bigger portion so that it's ready to use first thing in the morning. We use equal amounts by weight of water and flour to make a stiff batter, and by the time we use it, it needs to be ripe: bubbling, viscous, "talking" to us. If you use a starter infrequently, it needs to be rebuilt in stages, fed two or three times in succession (with the majority of it discarded). Again, the rhythm of

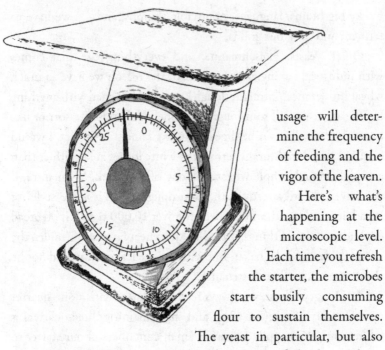

usage will determine the frequency of feeding and the vigor of the leaven.

Here's what's happening at the microscopic level. Each time you refresh the starter, the microbes start busily consuming flour to sustain themselves. The yeast in particular, but also the bacteria, produce carbon dioxide as one of the by-products of their metabolism. The tiny bubbles of carbon dioxide cause the starter to expand in its container. The bacteria are also producing the compounds responsible for the tangy flavors of sourdough, including lactic acid (also found in yogurt) and, to a lesser extent, acetic acid (also found in vinegar). After a while, as these organic acids and other by-products build up in the starter, they begin to inhibit the growth of the microbes. As a result, the expansion of the starter in the container reaches a plateau. After the plateau period, if no refreshment is forthcoming, the microbes become dormant and even begin to die. The starter continues to fall as carbon dioxide slowly escapes and is not replaced by further metabolism.

A relevant anecdote: Years ago, an experienced home baker came to me with a problem. Her starter of many years wouldn't function anymore. She was stumped! After chatting for a bit, it emerged that she was a professional engineer and had recently moved to a new home. Had she designed it herself? It must be energy efficient! Was it so "tight" it had an air exchanger? Yes . . . and that was the problem. No bacteria to feed off of. I suggested she take her starter out to the goat barn and give it a place to live out there, and it worked. Another friend in a brand-new house can't get his starter to grow in the winter, when the windows are all closed. Beware the over-hygienic abode!

How to Create a Starter Culture from Scratch

Creating a sourdough starter from scratch is not difficult. The microbes that you need are all around you: in the air, in your flour, on your hands. All you need to do is provide them with these three essentials: food, water, and shelter. Over the course of a week or so, a succession of wild yeast and bacteria will take up residence in their new home. Once these microbial populations reach an equilibrium, the starter will be ready for use as leaven for your bread.

Regarding the three essentials:

FOOD. The food is flour, preferably organic, stone-ground, and locally grown. The best flour for creating a starter is whole-grain rye, thanks to the multitude of enzymes. (You will need just over 1 pound [450 g] for the first week.) Once created, the starter can be maintained on a diet of unbleached white all-purpose flour mixed with a smaller amount of whole rye flour.

Water. The water should be free of chlorination. After all, the reason that municipal water is chlorinated is to limit the growth of microbes ("germs"). You can dechlorinate tap water by filtering it, by boiling it and allowing it to cool, or by simply leaving it out in an open bowl overnight. You can also buy a bottle of spring water and use that. Water can be at room temperature, except for the initial mix on Day 1, when it should be 100°F (38°C).

Shelter. Use a wide-mouth glass jar with a lid that fits loosely. It will be helpful to have two or three identical jars. Screw-top lids can be used as long as you don't screw them on tightly. Whenever you clean the jars, take care not to leave any soap or detergent residue. Your microbes will be quite comfortable in their new lodging if the ambient temperature is between 75 and 82°F (24–28°C) for the first 24 hours and in the range of 68 to 78°F (20–25.5°C) for the rest of the first week. Bear in mind that the starter will take longer to get going at temperatures in the lower portion of this temperature range.

Here is a recommended daily planner for creating a robust starter in one week:

Day 1. Measure 60 grams of warm (100°F [38°C]) water into a clean jar. Stir in 50 grams of whole rye flour and mix to a uniform consistency. Cover loosely. Place the jar in a warm spot, but not in direct sunlight, someplace where the temperature will hold fairly steady in the range of 75 to 82°F (24–28°C). Wait 24 hours. The mixture will rise a bit.

Days 2–3. Perform the first refreshment as follows: (a) Measure 60 grams of room-temperature water into a clean jar. (b) Stir the starter in yesterday's jar, making the froth collapse, and transfer 50 grams of starter into water in the new jar. (c) Stir in 50 grams of whole rye flour and mix to a uniform consistency. Cover loosely. (d) Keep the new jar in a space where the temperature is in the range of 68 to 78°F (20–25.5°C). (e) Discard the unused material from yesterday's jar. (f) Repeat on Day 3.

DAYS 4–6. The same as Days 2 and 3, except repeat every 12 hours instead of every 24 hours. At some point during Days 4 through 6, you will probably begin to see the culture rising and then falling over the course of each 12-hour refreshment cycle. If this does not happen by Day 6, simply extend the procedure for some additional days until the rise-and-fall cycle occurs. Hold off on the Day 7 procedure until then.

DAY 7 AND ONWARD. Measure 50 grams of room-temperature water into a clean jar. Stir the starter in yesterday's jar and transfer 10 grams into water in the new jar. Stir in 10 grams of whole rye flour plus 40 grams of unbleached white all-purpose flour. Mix to a uniform consistency. Cover loosely and store in a space where the temperature is in the range of 68 to 78°F (20–25.5°C). Discard the unused material from yesterday's jar. After 12 hours, repeat this refreshment procedure.

It's unlikely that a home baker bakes more than once a week, and one option is to store your starter in the fridge, then "wake it up" gradually over several days with feeds 12 hours apart. (A starter can be refrigerated for a week or two without being fed.)

Ever the contrarian, I believe any serious home baker who wants a vigorous starter should keep it in an old-fashioned crock on the counter and feed it every day. Like a pet! Think of it as culinary mindfulness training. That mess of flour and water and yeasts and cultures is a living thing, and so are you. If you want to be well fed, do the same for her: she's the Mother, after all. She's more resilient than you give her credit for, so if you miss a day, you will likely be forgiven. Keep trying.

The Boilerplate Formula and Beyond

up to my elbows
in last night's
proof

As described in the introduction, my path to craftsmanship was more than a little bumpy. I mentioned the glass-shard dough, but there was also the flour-cart-as-human-catapult event, the accidental no-salt loaves, the French pushpin debacle, and many others.

Five ovens, six mixers, and two grandchildren later, I have it a little more figured out. I can't tell you how many loaves that amounts to, because I don't count them. It's all the same loaf, new and ready, over and over again.

Perhaps it's due to my limited imagination, but if the truth be told, all of our bread recipes are just variations on a single boilerplate formula. I've never trained in someone else's bakery. One might call that "self-taught," but really the dough itself has taught me, and she does not suffer fools.

So, after 25 years of developing this particular style of artisan bread, perhaps I can share some details and techniques and spare you some of the grief I visited upon myself. Perhaps not.

To wit: use baker's percentages and the metric system for your recipes. This is the industry standard for delineating bread formulas. It's simple, it works, and thanks to the base 10 nature of the metric system, it can easily be scaled up or down. All ingredients should be weighed, in grams. Simply put, the baker's percentage system states that the total flour used is expressed as 100 percent, and any other ingredient is referred to in relation to that amount. To make it even simpler, our proportions of starter and salt remain more or less constant throughout the formulas. The water is also fairly constant, at least for the first mix of each dough.

More than 100 percent? Yes:

100% flour
75% water
12% starter
3% salt
100:75:12:3

It must be my Euclidean proof, my Pythagorean formula for this postmodern version of the Eleusinian Mysteries. The flour-water-starter-salt rule of 3's: 3 times 1, times 4, times 25, times 33.3333. Whatever it is, it works for the Hungry Ghost.

In sizing the formulas for home baking, there are two different approaches you can take: One is to use 1 kilogram of flour and easily calculate the weight of all the other ingredients from there. That will produce two very large loaves, just over 2 pounds (about 1 kg) each. The other strategy—one that will produce two standard 1½-pound (680 g) loaves—is to begin with 840 grams of flour. The table below outlines each scenario.

	Percentage	Two 2-pound (1 kg) loaves	Two 1½-pound (680 g) loaves
FLOUR	100%	1,000 g	840 g
WATER	75%	750 g	630 g
STARTER	12%	120 g	100 g
SALT	3%	30 g	25 g

That's my "formula." Yours now, if you want to use it. Of course, as the famous French baker Lionel Poilâne said, "The magic of bread baking is in the manipulation and the fermentation," so the formula by itself won't necessarily get you all the way there.

Let's break it down:

WATER: The most crucial source of life on Earth. Use clean water and, if chlorinated, filter it. Remarkably, sea water has exactly the right salinity for bread dough, but it is difficult to source, at least cleanly. Unlike flour or just about anything else, water always weighs the same, so you don't have to put it on a scale to know if you have the correct amount. The metric system is actually based on the qualities of water (0°C is when it freezes, 100°C is when it boils), and most conveniently for us, 1 liter of water will always weigh exactly 1 kilogram.

As constant as its weight might be, it is the temperature of water that varies widely and can sink or swim any baking project. It is one of the many variables that must be accounted for on any given day. Is your prep space warm? Is the tap water running cold? Even in our cozy bake house, we bring the water temps up in the winter and ice them down on hot summer days. Desired dough temperature is generally 78°F (25.5°C). You can average out the room and flour temps to determine the water temperature you need to achieve this.

We start with 75 percent hydration on almost all dough. That's high, according to the experts, but that's what we do. And it's up from there in the second mix (adding water during the second mix is called *bassinage* in French). Unless you're using a rather weak (low-protein) flour, this won't be a problem.

SALT: Almost any salt—fancy-schmancy, mined, or sun-dried from the sea—should do. Don't waste the large-crystal good stuff here, and avoid kosher salt, because it seems to have a lower salinity. We use salt from the Dead Sea, but not for any particular reason. Our proportion of salt is 3 percent. Again, that's somewhat high according to the experts, but I like to actually taste the bread, and the ionization of sodium and chloride on the human tongue is absolutely crucial. We also steer clear of iodized salt, because some customers have to avoid it for health reasons. If you're baking for folks who are watching their sodium intake, you may want to lower the amount to 2 percent.

STARTER: We use 12 percent natural leavening in all our formulas. That's low according to some, and doubtless high to others. Given the vigor of our starter, which is fed twice a day, plus the warmth of our working environment and the rhythm of our production schedule, that's what works just right for us. You can adjust up or down if you want a longer or shorter fermentation, if the leaven is sluggish, or if the room is cold. It is easier to adjust other variables, though, such as water temperature.

FLOUR: The real prima donna of any bakery. Protein levels, mill dates, levels of refinement—it's a lot to calibrate. Generally, flours are stored at room temperature, but they can certainly be refrigerated ahead of time on a hot summer day. Protein levels are a crude but useful scale of gluten strength, ranging from weak (pastry flour, 9 percent) to middling (all-purpose, 11 percent) to strong (bread flour, 12 percent) to very strong (pizza flour, 14 percent). Stronger is not always better, though, and most of our white flours are around 11.5 percent.

Whole and bolted wheat flours tend to be strong, and therefore "thirsty." It is tempting, but dangerous, to give them all that they seem to want. (See In Between "White" and "Whole Wheat" on page 38 for more on bolted flour.)

Khorasan is a type of durum flour originally from Iran (the northern region of Khorasan) and, like semolina durum, it is very strong and demands a high hydration—upwards of 83 percent.

Spelt is one of the trickiest flours to work with, as it contains more gliadin proteins than gluten. This makes it more extensible than it is elastic—it'll spread out, but it won't snap back! We use a blend of bolted and white spelt flours (40-60) along with a spelt starter. Many bakers prefer a loaf pan with this dough, but we coax it into a boule and it does quite well on the hearth.

Rye flour lacks gluten and has pentosans instead. They make the dough pretty sticky, but neither extensible nor elastic. That's why 100 percent rye breads, such as a German Roggenbrot, are made in a special loaf pan called a Pullman pan (which has a sliding lid and is sometimes baked overnight in a low oven). We blend our bolted rye 50-50 with a strong white flour and make a hearth round similar to our other breads.

In Between "White" and "Whole Wheat"

Some gave them white bread,
And some gave them brown;
Some gave them plum cake
and drummed them out of town.

In *Through the Looking-Glass*, it's the Lion and the Unicorn that battle through their diets. They represent the Conservatives versus the Liberals, the Scots versus the English, and it may go all the way back to the Wars of the Roses. Yet, it's high time to end this "flour fight." Here's my proposal for a peptic cease-fire.

Let's start with wheat berries, the basic building blocks of bread. What began as a wild grass became, over millennia of human manipulation, a domesticated grain. Seeds were selected, over and over again, across generations of farmers and crops, for the best yields, flavors, and use in batters, cereals, and feed. Civilizations rose and fell based on their relationship to wheat. The pyramids may be built out of stone, but they were carved out of grain.

A wheat berry is a tremendously well-designed packet of protein, very much like an egg. There's an outer shell of bran, a white starchy endosperm, and at its center, the germ. If stored in a cool, dry place, it is remarkably stable and can last a long time without spoiling. The gravest danger is generally vermin, hence the high status of cats in Egyptian mythology, protectors of the precious harvest.

Once the wheat berry is split open to make flour, it is suddenly vulnerable to degradation. As the oils in the germ are exposed to oxidization, the clock begins ticking: it'll all turn rancid sooner than later. Whole wheat flour is just that: *all* of the bran, endosperm, and germ, ground together, generally between two stones. It is full of vital enzymes and should be treated as a fresh ingredient and used as such. The ugly truth is that much of the whole wheat flour on store shelves is already rancid (or it contains a preservative such as benzoic acid), and your labor-intensive loaf has already been undermined. Either find a fresh local source (and read the mill date—under 2 weeks, at least) or get your own countertop grinder and discover the vibrancy and taste of the real thing.

White flour is made solely from the endosperm, the starchy stuff in between the germ and the bran. Because there's no germ or oils, there is no urgency to use it fresh. On the contrary, white flour needs to oxidize in order for it to be stable. That's why, in the bad old days, it was bromated and bleached. Sacrificed to the gods of Long Shelf Life. White flour is milled on a roller mill that slices the berry open, extracts the endosperm, and separates out the bran and the germ—two "by-products" that are then sold to second- ary markets. There's

nothing inherently wrong with white flour, though the fact that many governments (Canada and the UK, for example) still require flour companies to add back in essential vitamins (such as niacin, riboflavin, and calcium) that are abundantly present in whole wheat should tell you something. Bran is full of minerals, particularly iron (hence the reddish-brown color), which is why it can be bitter and why white flour is sweeter. The naturally occurring glucose is made more available in white flour, and this allows for the caramelization of the crust. We can't get a crunchy crust on a whole wheat loaf, and that is why.

In the old-fashioned Manichean view, there's a good versus bad, black-and-white dualism when it comes to wheat and white flour. I think this is terribly misguided and comically ahistorical. While it is true that white flour lacks certain nutritional components and whole wheat flour contains crucial fiber, the culinary truth is in the gray zone: not only do our taste buds crave the texture and loftiness of whiter loaves, but there are bran shards in whole wheat flour so large that they slice through gluten strands and collapse the structure of the dough.

The good news is that there's a win-win compromise, and it's not even newfangled. Even the Romans had the technology to sift flour and remove the large bran. They used a bolt of loose fabric to do it, and hence we have "bolted" flour. It retains all of the germ, all of the endosperm, and most, but not all, of the bran. There is in fact a wide range of bolted flours, from close to white to almost whole, and an entire confusing vocabulary that goes along with it, depending on whether you are French, English, or Italian. Some identify the extraction rate, some the ash content; they overlap but are quite distinct. That's not a wormhole we've got to go down, fortunately! Just discovering that there is a "whole-ish" wheat flour that provides both high nutrition and sweet, airy loaves is a revelation to most American bakers. The revival of regional mills (and grain growers!), plus the availability of affordable countertop mills, should spur the use of this refound bolted flour.

At Hungry Ghost, we are extremely fortunate to have the nearby Ground Up Grain deliver every Friday (in their cool electric van), and to receive the bounty of freshly ground and bolted rye, spelt, pastry, pizza, and bread flours. We use bran, mids (short for middlings, a by-product of the bolting process), malted barley seeds, and barley flour from them as well.

We use these flours every day: our Eight-Grain and Hat Trick loaves are made entirely from Ground Up's bolted bread flour, as is our Beet & Coriander Fougasse. Our pizza dough (sold frozen) is also entirely made with their bolted pizza flour, and it is remarkably extensible and tasty. The bolted rye goes into both our Rye and Country loaves, and the bolted spelt, not surprisingly, into the Spelt Bread. Bolted pastry flour is a key ingredient in our ever-popular Buttermilk Biscuits. We use the bran daily in our Eight-Grain mix and in our weekly bran starter (for Colonel Buckwheat Bread), and the mids we blend with rice flour to dust the Eight-Grain loaves.

When Hungry Ghost first opened in 2004, there was no way to source regional wheat. At all. I did call the Massachusetts State Agricultural Commissioner's office one day way back then and had the following conversation:

"Can you tell me if there are any farmers in the state growing wheat?"

"Wheat? What a crazy question! What do you want that for?"

"Well, we're bakers, so it's kind of necessary for the process."

"You probably wear jeans, too, so you might as well search for cotton farmers! Goodbye!"

Little did that bureaucrat know, the first wheat grown in the United States was right off Cape Cod Bay in 1602. Since that call, many organizations and individuals, including Maine Grains and the Northern Grain Growers Association at the University of Vermont, have helped revive the regional grain economy. We still have a long way to go, but having a local mill is a huge step!

I always keep some show-and-tell jars handy to show the curious what wheat berries—and bran, and germ, and white or whole wheat flour—look like. We often have wheat growing out on the lawn or in a planter outside so we can all be reminded that it's really a grass! The source of our sustenance should not be a mystery. The awe comes from actually knowing.

Starters, Dusters, and Offbeat Inputs

By and large, bread is (or should be) made with as few ingredients as possible: flour, water, starter, and salt. And while most of us eschew the more expensive luxuries of chocolate, cherries, Asiago, walnuts, and other goodies to fold into the dough, there are countless other more modest ingredients that can distinguish a basic loaf.

Varying the flours, of course, is the first and most consequential option: whole, bolted, or white wheat flours, for instance. Blending in a portion of semolina (the hard, yellow, high-protein flour from durum wheat, most commonly used in pasta) or one of the older, less-domesticated varieties of "ancient" wheat, such as spelt or khorasan, is also easily done. Rye flour is particularly strong in flavor but weak in protein.

Once the base flours are established, what kind of taste notes do you want in a loaf? Sweet, savory, herbal, earthy? Many of the standard add-ins are very dependable (and affordable): caraway, sesame, pumpkin, or sunflower seeds. Rosemary, fennel, sage, and basil leaves.

We use chamomile flowers in our Spelt Bread, sage in our Fig & Sage Bread, and dried onion in our Olive-Semolina Fougasse.

We assiduously avoid using non-glutenous flours, such as amaranth or chestnut flour. Why bother? If you use just a little bit, it will get lost in the dough and you won't taste it. If you use too much, it'll weigh down the crumb. If there's a grain worth trying, don't turn it into a fine powder! Use it in other forms for texture and taste. Barley, for instance—a terrific flour for cookies, but not for bread. Instead, we put whole malted barley seeds directly into our Hat Trick dough. It provides a sweet and crunchy contrast to the bolted wheat crumb that surrounds it. Oats? We dust the same Hat Trick loaf with raw oat flakes by dunking the still-sticky, just-shaped rounds into a bowl of oats before dropping them into their overnight baskets. The oats toast on the outside of the loaf while baking and add their own distinctive sweetness.

Country Bread has a sort of parallel but inverted structure: stout-soaked rye flakes (like oats, only grayer) go into the dough (a blend of white and rye), then the rounds get a coat of sesame seeds before basketing.

Cornmeal Dance Bread has beer-soaked corn grits in a rye-white dough blend and a crust of sunflower seeds.

Colonel Buckwheat Bread, on the other hand, has a cornmeal crust with toasted buckwheat groats in the crumb. On the Raisin Bread, there's a coat of cocoa powder cut with a little white rice flour: dark, yummy, and mysterious. The Spelt loaf is dusted with quinoa flakes that not only keep it from sticking to the basket, but also caramelize in a unique, subtle way. The name of the Fig & Sage Bread identifies this loaf's interior ingredients (which are paired with semolina blend), and it has the added bonus of a flax and millet crust.

And then there's our Wheat Germ Theory Bread, which reconstructs the original wheat seed with toasted wheat germ inside the white flour dough, along with a bran-dusted crust. This is what the French might call an integral bread: not white, not whole, but somewhere in between.

Bran gets used in a number of ways by the Ghost: six days a week it makes up a hearty proportion of our Eight-Grain soaker that goes into the loaf of that same name. Equal amounts of sesame, flax, and sunflower seeds, along with millet, oats, rye flakes, and corn grits, are mixed with bran, water, and discard starter. They soak overnight before being added to the dough. The fermentation vastly increases its digestibility.

Likewise, we use a 100 percent bran starter for our Colonel Buckwheat dough. As a leaven, it looks nothing like regular sourdough starter. No viscosity, no bubbling, not even much scent. But it pushes the dough just as well as any other starter, and of course it also reintegrates the fiber and minerals back in with the white flour.

One last wacky but delicious experiment is the blue cornmeal starter: it smells like bubble gum! It leavens our Raisin Bread and our Fig & Sage Bread more than adequately while providing a complex (and guilt-free) sweetness.

Baking and Breaking Bread

remember, an
apron's just a
backwards cape

Scale aside (from the weight of puns!), home baking and bakehouse work differ in many ways: There are systems in any bakery that are tough to replicate at home. When I go to work, I am in certain ways stepping into a stream of dough and just catching the fish. At home, you've got to collect rainwater and build a hatchery and sew your own nets. Our walk-in cooler at Hungry Ghost is set to 42°F (5.5°C), an optimal temperature for retarding dough. Your refrigerator is quite a bit colder than that. We're able to crank our oven heat while your (perhaps rental) unit's dial is likely uncalibrated. Of course, you're only making a couple of loaves at a time, and presumably the pressure is off: they'll not be on display or even for sale. But bread is bread, and our goals are the same, even if the volume's turned way down. There is no shame in being an amateur: the word means *lover of . . .*

Mixing the Dough

As a home baker, you can mix either by hand or by means of a stand mixer. The formulas assume use of a mechanical mixer, but they can all be followed without. Either way, mixing sourdough always has two phases. The purpose of the first phase is simply to incorporate flour, water, and starter together. The purpose of the second phase is to develop the dough so that the gluten will be able to trap gas bubbles and expand, or rise. The second phase is also the time to add salt, which helps the dough develop, and to add any other ingredients that will be included in the dough.

The main advantage of mixing by hand is that you have direct contact with the dough. The more bread you bake, the more you will learn how to "read" the dough. Does the particular flour that you're using today need a little more water to reach the right consistency? Has the dough been mixed long enough that the gluten is well developed, so that it will rise nicely and not collapse? Over time, you'll develop the skill to answer these and other questions about your dough.

If you're just learning to bake sourdough bread, I recommend starting out mixing by hand. That way you'll learn more quickly how to read the dough. Later, if you opt to use a stand mixer, you'll know what to look for when you stop the mixer from time to time to check the feel of the dough.

How to Mix by Hand

For the first phase of mixing, you'll use two types of hand movements: first paddling, then turning the dough. For each of these movements, keep your fingers held together, as in a mitten, while using your thumb as specified for that movement. Use one hand to mix the dough and use the other hand, which you keep clean, to rotate the bowl.

FIRST MIX. Disperse the starter in the water. Stir briskly, using your hand like a paddle. Break up the starter into smaller bits suspended in the water. Add the flour and continue stirring until all the flour is wet, although still somewhat lumpy. (Some recipes call for additional ingredients, like honey or olive oil, to be added at this time.) It's desirable to stir always in one direction, either clockwise or counterclockwise.

AUTOLYSE. For most of the breads in this book, the next phase is the autolyse. Cover the mixing bowl with something to prevent the surface of the dough from drying out. You can use plastic wrap, a silicone stretch cover, a shower cap, a beeswax wrap, or a wet tea towel that has been wrung out. Then place the bowl in a warm spot—ideally 78 to 80°F (25.5 to 26.5°C)—for 20 minutes. (If this time interval stretches to 30 minutes, or even a bit longer, there's no problem.)

SECOND MIX. Now you will turn the dough. Wet your mixing hand. With your four fingers again held together as a paddle, slide your hand between the bowl and the underside of the dough. Clamp down on the top with your thumb. Lift the dough away from the edge of the bowl, then flip the outstretched flap of dough downward and across the center of the dough mass, repeating as you make your way around the perimeter of the bowl and using your clean hand to rotate the bowl. During this step, you can wet your mixing hand repeatedly to reduce the tendency of the dough to stick to your skin. This is also a nice way to add some extra water (known as *bassinage*) to the dough.

How long should you continue the second mix? The answer will depend on your skill and your pace. In general, 10 to 15 minutes is typical. When mixing by hand, it's better to mix for longer rather than shorter if you're unsure. You won't hurt the dough if you mix more than necessary—it's almost impossible to overmix bread dough by hand. As your ability to read the dough improves over time, you'll be able to better gauge how soon you can end the second mix.

Most of the recipes in this book call for the salt to be added partway through the second mix. To add salt, first generously wet your mixing hand, spread your fingers, and stipple (pinch) the surface of the dough with your fingertips to get a dappled texture. Then, with your dry hand, sprinkle about a third of the total amount of salt across the dough surface. Resume turning the dough for a minute or so to get the salt mixed in. Repeat this sequence twice more: wet the hand, stipple the dough, add some salt, and turn the dough. You'll want to start adding the salt at least 3 minutes before you expect to finish mixing so that it all gets mixed into the dough. If you start after 6 or 7 minutes of mixing, you'll be sure to get it all in, even if your overall mixing time turns out to be just 10 minutes.

If you want, you can take a couple of short breaks during the second mix, like 2 or 3 minutes each, or one break of about 5 minutes. Just cover the dough so that the surface will not start to dry out. You'll notice that the dough will relax slightly during the rest period, making it easier to work with when you start in again.

How to Mix with a Stand Mixer

At Hungry Ghost, we mix our ingredients in a big spiral floor mixer that can handle up to 100 pounds (45 kg) of dough. Purpose-built for bread-making, it has just two speeds. The first phase of mixing takes place on first speed (slow) and the next on second speed (faster). Stand mixers for home use, on the other hand, are multipurpose machines with as many as 10 or 12 speed settings and several different attachments for mixing. The models most often purchased have a tilting-head design and come with a C-shaped dough hook along with a flat beater, or paddle attachment. These instructions are keyed to those models.

FIRST MIX. To prepare for stand mixing, unlock the mixer head and tilt it back. Use the paddle, not the dough hook, for the first mix. Put the water, starter, and flour into the bowl. (Some recipes call for additional ingredients, like honey or olive oil, to be added at this time.) Tilt the mixer head down and lock it in place. Run the mixer at its lowest speed for 4 minutes, stopping once the water, starter, and flour have been incorporated into a shaggy mass.

AUTOLYSE. For most of the breads in this book, the next phase is the autolyse. Unlock the tilt mechanism and tilt the mixer head up. Remove the bowl from the mixer and cover it with something to prevent the surface of the dough from drying out. You can use plastic wrap, a silicone stretch cover, a shower cap, a beeswax wrap, or a wet tea towel that has been wrung out. Then place the bowl in a warm spot—ideally 78 to 80°F (25.5 to 26.5°C)—for 20 minutes. (If this time interval stretches to 30 minutes, or even a bit longer, there's no problem.)

SECOND MIX. Replace the paddle attachment with the dough hook. The first time you make a particular recipe, take the extra step of oiling the surface of the dough hook with olive oil, including the dough guard near the top of the hook. Oiling the hook helps prevent the phenomenon of "dough climb," which is described below. If you find, with experience, that your mixer does not have problems with dough climb, then oiling the hook is not needed.

Return the bowl containing the dough to the mixer stand, lower and lock the tilting head, and mix at whichever speed setting the manufacturer recommends for bread dough. For most mixers, this is speed setting #2. If the manufacturer recommends speed setting #2 or #3, then use #3. Continue mixing for 3 minutes. Before long, you should see and hear the whirling mass of bread dough throwing arms of dough outward to hit the wall of the bowl.

Dough climb is a common occurrence at this point. The dough winds its way up the dough hook toward the dough guard, wrapping itself tightly around the hook so that the mass of dough simply spins without contacting the walls. When that happens, no mixing is taking place. Although the dough guard is intended to prevent the dough from climbing further, dough can sometimes clamber right over the guard and beyond, up the mixer shaft and into the motor housing. There are two ways to address this problem. One is to stop the mixer, slide the dough off the hook, add 2 teaspoons of water to the bowl, and re-oil the hook. The other is simply to raise the speed by one speed setting, thus increasing the centrifugal force on the dough so as to fling some of it off the hook. (Some manufacturers caution against mixing dough at a speed above that recommended in the manual. Fortunately, high-hydration doughs such as those in this book offer relatively low resistance to mixing. So you're not likely to overload the mixer's motor by running it for a few minutes at one or even two speed settings above the recommended setting. But if you do observe that the motor housing is overheating, shut off the mixer and allow it to cool.)

Most of the recipes in this book call for the salt to be added partway through the second mix. Turn off the mixer, unlock the tilt head, and raise it. Slide the dough off the hook. Add the salt to the dough, along with 2 teaspoons of water. Re-oil the dough hook, lower and lock the tilt head, and restart the mixer at the original second mix speed. Again you should see and hear the whirling mass of bread dough throwing arms of dough outward to hit the wall of the bowl. As the process continues, you'll see that the dough no longer sticks where it hits the wall. After a total of 5 minutes of the second mix, if there are no other ingredients to add, mixing is done. If the mixer was running at a speed setting higher than what the manufacturer recommends, then a total of 4 or 4½ minutes for the second mix may suffice.

A word of caution: do not leave the mixer unattended while it is running. Many stand mixers tend to jump a bit when mixing dough and may even "walk" right across the counter as a result. So stand by your mixer and steady it with your hand if necessary. Also, you'll want to know when to intervene if dough climb starts to develop.

Bulk Fermentation

Bulk fermentation is the phase when most of the fermentation occurs, as yeast and bacteria from the sourdough starter multiply and metabolize the carbohydrates in the flour. Bulk fermentation officially begins when mixing ends, although actually the microbes begin metabolizing and multiplying as soon as the starter comes into contact with the flour and water at the start of mixing. The rate at which they metabolize and multiply depends on the temperature.

For most of the breads in this book, the criterion for ending bulk fermentation is a twofold increase in the volume of the dough. That is why it's so helpful to ferment in clear or translucent containers with straight sides. You can mark the level of dough at the start of bulk fermentation with a marker or a piece of tape, then observe over the next several hours until the dough doubles in volume. Be sure to oil your container well before putting the dough in.

As the dough expands during bulk fermentation, you'll observe several changes besides the volume:

- The contour of the dough will become convex.
- The surface texture will become smooth with small bubbles. Bubbles inside the dough will also be visible along the walls of a transparent container.
- To the touch, the dough will shift from stickiness to tackiness; it will cling more to itself than to the container wall or your hand.
- The consistency will become soft and airy. Words that may occur to you include "pillowy" or "marshmallowy."
- When you stretch the dough as part of the folding process, you'll notice more resistance to stretch and less extensibility.

These changes are helpful as auxiliary clues, especially if it's difficult for you to gauge volume expansion accurately (if your dough is in a bowl, for example, rather than in a straight-sided clear container).

Note that the criterion for ending bulk fermentation is a specified degree of expansion of the dough, rather than a defined time interval. My Hungry Ghost recipes do include an estimated time, such as "5 or 6 hours." The purpose of these estimates is to help you plan your baking days. But, as the saying goes, "Your mileage may vary." Two of the main sources of variability are the differences between my bakery and your kitchen with respect to the amount of dough and starter management. If you understand these sources of variability, then you can improve the predictability of your bulk fermentation times.

Bear in mind, again, that the time needed for bulk fermentation depends on temperature. There are two temperatures that determine the time needed for bulk fermentation: the dough temperature at the start of bulk fermentation (final dough temperature) and the air temperature while bulk fermentation takes place (ambient temperature). If the ambient temperature differs from the dough temperature, then the dough temperature will drift toward the ambient temperature as bulk fermentation proceeds. The metabolic activity of bacteria and yeast does generate a small amount of heat within the dough, but the quantity of heat produced is small, especially during the earlier hours of bulk fermentation.

Fermentation time in the home kitchen is more sensitive to air temperature than in the bakery. Why is that? For purposes of this discussion, let's say your desired dough temperature is 78°F (25.5°C). Suppose that, today, the temperature is 68°F (20°C) inside the bakery and also in your kitchen. I mix 100 pounds (45 kg) of dough in the bakery with a final dough temperature of 78°F (25.5°C). Meanwhile, you mix 3½ pounds (1.6 kg) of an identical dough at the same temperature. Starting at the same time, we each let our dough ferment for 5 or 6 hours at room temperature and then recheck the dough temperature. The small dough in your kitchen has a much lower thermal mass, so it cools down more and takes longer to ferment. Furthermore, if your dough temperature is below target at the start of fermentation *and* the air temperature is low during fermentation, then the delay will be compounded.

In this example, how would you improve the consistency of your results, so that your bulk fermentation time is close to the time estimated in the recipe?

- Try to land the dough temperature at the end of mixing as close to 78°F (25.5°C) as you can. How? When mixing the dough, calculate the optimal water temperature as described in the boilerplate formula (see page 35). Warm or cool the water to the optimal temperature before mixing.
- Ferment the dough in a spot where the temperature is as close to 78°F (25.5°C) as possible. If the temperature in that spot is lower than 78°F (25.5°C), you can compensate by using extra-warm water to start the fermentation somewhat above 78°F (25.5°C), with the understanding that dough temperature will fall below 78°F (25.5°C) sometime during fermentation. That way, the average fermentation temperature can be somewhere close to 78°F (25.5°C).
- If you undershoot and the initial dough temperature is well below 78°F (25.5°C), the only way to compensate is to run the fermentation at an ambient temperature above 78°F (25.5°C).
- If you can't limit or balance out temperature deviations, you can still make good bread. You just need to adjust the bulk fermentation time.

Over the course of the bulk fermentation period, you'll need to fold the dough several times. The number of folds needed will depend partly on how far the dough was developed during the mixing process. Four or five folds, spread fairly evenly over the course of bulk fermentation, will typically suffice for the breads in this book. In your early work with sourdough, I recommend folding the dough every hour or so. The envelope fold is a straightforward gesture that involves pulling each of the four sides of a binned dough over the top of itself, to stretch and gently rearrange the developing gluten structure.

As mentioned above, the physical properties of the dough will change as bulk fermentation continues. You'll notice that as you perform each successive fold, the dough offers somewhat more resistance to stretching. Also, as you complete each successive fold, you'll notice that the dough tends to form more of a pile in the center of the bin. The dough will then relax and spread out some during the interval before the next fold. But, if the folds are spread evenly over time, you'll also notice that the dough relaxes less after each successive fold. These are all signs that the dough is developing.

As your folding technique improves over time, together with your ability to read the dough, you may find that folding less often still yields excellent results. With experience, you may find that three folds will suffice; or some of your dough may need six. If you're unsure, it's better to fold more often rather than less. You won't hurt the dough if you fold more times than necessary. In general, though, the last fold should be done at least 30 to 60 minutes before the end of bulk fermentation and it should be done very gently. That way you'll avoid de-gassing the dough.

Dividing

Bulk fermentation ends when we divide the dough into smaller pieces, each of which will become a loaf. At Hungry Ghost, we're dividing a large dough mass into many pieces. We weigh each piece to ensure uniformity. As a home baker, such precision is not necessary, but you should still try to divide the dough into equal parts so that the pieces will bake evenly. The bread recipes in this book each yield two loaves. Before dividing, you may find it helpful to draw the dough knife very lightly once across the dough where you plan to cut. Then take a moment to eyeball the two sides to make sure they look equal before you cut. As you gain experience, you won't need this extra step.

Shaping Super-Slack Dough

The driving principle behind the bread at Hungry Ghost is pushing the envelope in terms of hydration, fermentation, and gelatinization. That means more water in the dough while mixing, more bubbles in the dough while shaping, and more time in a hotter oven when baking.

A high-hydration dough is called *slack*, and it is a challenge to manipulate. Many a bakery visitor (professional or not) has tried to shape our dough and found it unmanageable. Our formulas begin at 75 percent water and go up from there. After being folded a number of times throughout the day, the doughs have good structure, but they are certainly sloppy and sticky.

One way to describe our approach is to handle the dough without actually touching it too much. Or at least, never too long in any one spot. Our hands have grown accustomed to spinning the dough in place on the work surface, first across the outer palm and then each fingertip in quick succession.

Start with a clean, scraped, and flour-free section of a wooden table. Take your hunk of dough and fold it in on itself from at least 3 or 4 points, then flip it over. Let the messy underside start to stick to the table a little while you gently turn it into a round. The table is your third hand, an anchor allowing you to create tension in the skin of the dough and stretch it smooth. Imagine that you're winding a coiled spring. Don't lift it off the table and don't douse it with flour (though your fingers might use just a touch). Tuck it down under and turn. Doesn't matter whether it's clockwise or counterclockwise, as long as it's in one consistent direction. Rather than grabbing and overhandling it, coax it into a sphere, a little planet of dough.

If it falls apart or blows out like a burst car tire, lay it aside for a few minutes. The gluten strands are just like muscles that have tightened up too much, charley-horsed. They'll relax after a little while. Try again, but . . .

Know when to stop! When it looks like one of Jupiter's moons, scrape it off the table with a clean dough knife, make sure the seam, where the belly button of the dough might be, is closed up and, if not, seal it with your hand. Then slather it with flour or whatever duster you're using and lay it in an equally dusted basket. Add a little more duster around the edges so that it won't stick to the sides of the basket as it expands. You'll have plenty to brush off when it's time to bake, but nothing's worse than a ripped and deflated "could've-been" loaf.

Proofing

After the bulk proof development and the subsequent shaping of each loaf, there is a secondary proofing stage during which each dough rises in its individual basket.

The real modern technological break-through in artisan baking has not been new ovens or fancy mixers or hybridized wheat. It's refrigeration. With affordable walk-in

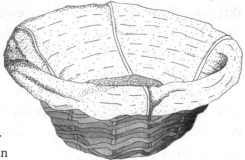

coolers, bakers can now engage in slow, overnight dough fermentation *and* get a good night's sleep! It's truly a win-win situation, with happier bakers and better-tasting bread to show for it.

By cooling the dough and slowing the fermentation *way* down, known as retarding, not only can we bakers sneak away, but the dough can develop more complex lactic and acetic acids. How cool, for how long? Well, that's the question that only repeated trials and good luck can answer. This is where fine-tuning your production rhythm comes into play. A colder dough can take more time; a warmer one, less. Judging where you want that dough to be when you get back to it is like pitching a boomerang and knowing where to be to catch it.

Our walk-in cooler is set at 42°F (5.5°C). If the dough itself were to reach 41°F (5°C), all fermentation would cease. It would be akin to cryogenics. But we don't want the dough to be suspended indefinitely; we want it to have an interesting dream-life. Your refrigerator will be a little colder, because the FDA and the CDC advise that refrigerators be kept at or below 40°F (4.5°C), so you'll need to take that into consideration. When warm dough goes into the refrigerator, it continues to ferment as it cools, albeit slowly. With a starting temperature of 78°F (25.5°C), the center of a typical round loaf reaches 65°F (18.5°C) after about 3 hours in our fridge, 50°F (10°C) after about 7 hours, and 42°F (5.5°C) after about 12 hours. Once the dough reaches that low temperature, any additional time is basically holding the dough until it's a convenient time to bake.

In the bakery, our doughs are retarded for 15 to 24 hours. That's a long trajectory. At home, 12 to 18 hours is appropriate. Your lower fridge temp means there's less chance of overproofing, but also more time needed (1 or 2 hours) to bring the dough up to room temperature before baking.

Baking

At Hungry Ghost, the real test comes the morning after all that mixing, proofing, shaping, and slow-fermenting. Contrary to common practice (again!), we transfer dough directly from the cooler into the oven. Consequently, our oven temps are a little higher. The dough, though slack, is much easier to remove from its basket while cold and is more effectively scored with the baker's blade.

You need to hurry but not be frenetic when brushing off the extra flour dust (or seeds or cornmeal), giving each loaf its distinctive slash, and hustling each boule or bâtard into its steamy spot. As ever, the dough is the boss: if the boules are underproofed, give them more floor time to let them warm up and soften. If the bâtards are browning up quickly in the oven, pull them out as soon as they're ready, regardless of what the timer says. Each formula in this book offers additional instruction, some of it repetitive, and some perhaps picayune, but all in the interest of sharing experience and saving you some grief.

We bake up to 65 loaves at a time in our big wood-fired, steam-injected masonry oven. Here's how to adapt our process for your home oven, whether gas or electric.

First, take your dough out of the fridge and uncover it so the dough has time to come up to room temperature. Preheat the oven, with your steam setup inside, for at least 45 minutes. Make sure you have a supplemental thermometer to assess the actual temperature of the oven. As discussed in Tools of the Trade, the preferred steam setup is a Dutch oven used as a baking vessel, and the alternative setup is a steam pan. If you're using a steam pan, it goes on the rack below the rack holding the bread. In this case, put a baking stone or parchment-lined sheet pan on the rack above the steam pan.

To prepare for loading the oven, dust the top surface of the dough (which will become the bottom of the loaf) with semolina or with a 50-50 mixture of all-purpose flour and white rice flour. Gently transfer the dough, dusted-side down, to the preheated Dutch oven, if you're using one, and score the top of it decisively. Cover and put it into the hot oven. Or, if you're using a steam pan, put the dough on a peel, score the top, and transfer it to the preheated baking stone or sheet pan. Then give thanks to Hestia, goddess of the hearth.

The Loaves

scoring Pi
on the Country
an infinite number of times

The sixteen recipes for loaf breads follow a thematic and also developmental sequence, introducing new techniques and classes of ingredients in a progressive manner. But you don't need to bake them in the order that they appear—feel free to hop, skip, and jump around!

French Bâtard is a simple classic, made of just flour, water, salt, and leavening. This bread is a great starting point because you can focus on the basic bread-making ingredients and techniques before moving on to more complex recipes.

Rosemary Bread introduces the round loaf. It also goes beyond the core formula of flour-water-salt-leavening by flavoring the dough, in this case with herbs. *Semolina-Fennel Bread*, another herbed bread, also ushers in semolina, a distinctive type of wheat flour that plays a supporting role in various breads at Hungry Ghost.

Think of the next two breads as characters from "The Miller's Tale." *Hat Trick Bread* introduces bolted flour, which is neither white flour nor whole wheat, but is milled in between. It also is the first multigrain bread in the book and the first with a crust finish, a coating added to the outside of the loaf. *Wheat Germ Theory Bread* is a recipe that unpacks the wheat kernel into three distinct components, using wheat germ, wheat bran, and white flour as separate elements in the bread.

The next group of breads each contain a soaker, a grain that has been soaked in water or another liquid for a period of time before it is added to the dough. These are *Eight-Grain Bread, Colonel Buckwheat Bread, Cornmeal Dance Bread*, and *Country Bread*.

Fig & Sage Bread and *Raisin Bread*, sister fruit breads based on the same blend of two flours (white bread flour and semolina flour), are almost dessert breads. They both pair well with cheese.

The next three breads bring to center stage one grain that has stood in the background so far and two others that have not yet played a part: *Rye Bread, Khorasan Bread with Pumpkin Seeds*, and *Spelt Bread*. You will become familiar with the unique handling properties of each of these flours. The latter two have gained in prominence and popularity in recent years, as they are ancient grains.

The bread recipes conclude with two festive breads. At Hungry Ghost, we make the *Cranberry Holiday Bread* for Thanksgiving and Christmas each year. We bake *Challah* on Friday afternoons, before the Jewish Sabbath. Both of these are in the category of enriched breads, meaning that the dough includes sweetener (honey), oil, and egg. Challah introduces another shaping technique thanks to its familiar braid, as well as a different type of crust finish: egg wash.

French Bâtard

Formula
MAKES 2 (1½-POUND) LOAVES

Water	75%	630 g
Sourdough starter	12%	100 g
All-purpose flour	80%	672 g
White bread flour	20%	168 g
Salt	3%	25 g

50-50 all-purpose flour and white rice flour mix, for dusting

---

A French *what?* Bâtard! Yes, it means just what it sounds like! The "love child" of a baguette and a boule—squatter than the first and longer than the second. Don't we ever get tired of customers mispronouncing it, whether by accident or not? Sure, but it beats competing with everyone's expectations of what a French baguette ought to be.

Raised in a bilingual milieu, I'm immune to the notion that anything French is inherently more elegant. Baguettes, for one: they are a *thing*, to be sure, but they're not *my* thing. Baguettes are not, by definition, naturally leavened. They are straight doughs, made with yeast. Because of this and their large surface area, they go stale very quickly. Contrary to popular imagination, they are not that traditional or historic, either. They are a creation, like so many wild and wonderful things, of the Belle Époque. Lacking a black beret, a view of the Eiffel Tower, and the makings of a croque monsieur, I will just do without.

So, our French bread is bastardized: plumper, and naturally leavened. It is more digestible and has a much longer shelf life to boot.

Still crusty, still the perfect utensil for sopping up and cleaning plates before they leave the table. Make sure you've got a sharp-enough serrated knife. Cue the Django Reinhardt and skip the Gauloises.

The secret, if there is one, to successful French bread is to not use exclusively bread flour, but rather mostly all-purpose flour. Bread flour by itself is too strong and the crumb will be too dense and even. For most of our breads, but this one in particular, we need a somewhat weaker flour that will give us a slacker dough and uneven bubbles in the crumb. One should also be wary of any additives to the flour, and I don't just mean evil bleaching agents and bromate. These were added to whiten up the product and to oxygenate it. Ironically, white flour is the opposite of whole wheat in this manner: while you want the latter to be as fresh as possible, the former should be weeks or even months away from its milling date. Anything less than that can be considered "green" flour and may be unstable, particularly for long-fermented dough. Likewise, any natural additives—such as vitamin C, barley malt, or fava bean powder—that will assist a straight-dough baguette can actually interfere with the progress of our hero, the sourdough French bâtard. Some white flours are fortified with niacin, riboflavin, calcium, or folic acid to compensate for all those minerals taken out with the bran and germ; these are quaint WWII-era nutritional guidelines that have somehow survived. Best to get flour that contains only wheat. No conditioners, stabilizers, or enhancers. They are as icky as they sound!

No cigarette ash, no Seine water, no Montparnasse paint drops here: just water, flour, salt, and leavening. *Tout juste*. It is simple, but requires attention. Hard to get right, but perfect when it lands. Fun to shape and bake, but excruciating when it sticks to the couche or flops off the peel. Practicing French bread is like practicing French:

any Parisian feels free to correct you, but gradually your vocabulary expands and the conversation can take flight. The grammar of this dough can be elusive and its conjugations aggravating. But when the cheese and wine finally arrive, everyone will cheer the baker.

Instructions

Combine the water, starter, and flours in a mixer fitted with the paddle attachment. Mix on first speed for 4 minutes, then cover and let it rest in a warm place for 20 minutes.

Replace the paddle attachment with the dough hook; oil the hook if necessary (see How to Mix with a Stand Mixer, page 48). Uncover the dough and mix on second speed for 4 minutes, adding the salt and a little water after 2 minutes. Pinch the dough to get a sense of its slackness or tightness. Add more water if needed so it's not too sloppy, but not too stiff. Steer into the wind a bit, but don't let the sail luff. Added water is like the wind, and you are trimming the dough toward a distant point on the horizon....

Park it in an oiled bin, cover, and place it in a warm spot until doubled in size, 5 to 6 hours, giving it an envelope fold every hour or two.

Divide the dough in half and practice rolling an elongated loaf. It can be fat in the middle and tapered on each end, or more even all the way along. Try to use the whole length of your outstretched hands, pinky to pinky, as if you were reaching for two octaves on a piano. Slap the hunk of dough down and roll it in on itself, as evenly as possible. Avoid the dreaded dog-bone effect—skinny in the middle and bulging out on each end! Keep your workspace free of extra flour so that you can rock the emerging bâtard back and forth, seam down, pulling with your fingertips and pushing with the heels of your hands. This will create surface tension and seal the bottom when done with a confident motion. The dough is the boss, but you are the pushy assistant: you know where the next meeting is and what they need to wear. Get them into shape and move them along with conviction.

Instead of baskets, we use couches with our bâtards—long pieces of cloth, usually muslin or heavy cotton. Use a blend of white rice flour and white flour on the cloth, generously and evenly sprinkled, and then make little raised walls in between each bâtard. Allow for the fact that the loaves will expand, so don't squish them in too tightly or with too little material in between. Dust them enough to keep them from sticking, but not so much that the resultant loaf will have unincorporated flour on the bottom. Put the couched loaves on a small sheet pan, tie a food-grade bag over it all, and place it in the fridge for the night.

The next day, bring the loaves up to room temperature. Preheat the oven to 500°F (260°C), with your steam setup and baking stone or parchment-lined baking sheet inside. Slash the loaves at a 45-degree angle three times, parallel (*liberté, égalité, fraternité*!), and use a peel to put them on the baking stone. Turn the oven down to 450°F (232°C) immediately.

Uncork a nice red Buzet and put on some Sainte-Colombe viola da gamba music. After 40 minutes, take the loaf out and tap the bottom. If it's hollow sounding, let it cool on a rack and unwrap some hard chèvre cheese. *Vive la révolution!*

Rosemary Bread

Formula
MAKES 2 (1½-POUND) LOAVES

Water	75%	630 g
Olive oil	7%	59 g
Sourdough starter	12%	100 g
White bread flour	100%	840 g
Salt	3%	25 g
Dried whole leaf rosemary	1.4%	12 g
50-50 all-purpose flour and white rice flour mix, for dusting		

This is easily the most romantic bread we make. We see it seduce customers with its perfume and fly off the shelves, oven-warm. The Romans carried rosemary to Britain, and Charlemagne had it planted at every monastery. This herb beguiled Pliny and Cervantes. And Simon and Garfunkel, too. So why not you?

It's almost just a French round with herbs and some olive oil, but romance is always more than the sum of its parts. Something in rosemary's fat-soluble camphors, phenols, and carnosols plays pinball with one's lower sinuses, in a sexy kind of way. Not always so bad, being led by the nose . . .

Anyhow, that's Rosemary Bread. Popular but proper, it gets along with most anyone, from cream cheese to strawberry jam to squash soup! Conjure her up, invite her in, and you will happily make room for this unselfconscious diva.

Instructions

Combine the water, olive oil, starter, and flour in a mixer fitted with the paddle attachment. Mix on first speed for 4 minutes, adding however much water seems necessary to keep it from binding together too tightly. Cover and let rest in a warm place for 20 minutes.

Replace the paddle attachment with the dough hook; oil the hook if necessary (see How to Mix with a Stand Mixer, page 48). Uncover the dough and mix on second speed for 4 minutes, adding the salt and a little water halfway through and then the rosemary for the last full minute. Add more water, if needed, to make it a floppy (but not sloppy) dough.

Transfer to an oiled bin and cover. Bulk proof in a warm spot until it doubles in volume, 4 to 6 hours, giving it envelope folds every hour or so. Divide the dough in half and shape each into a nice, tight round, dust well, then place in proofing baskets. Store, covered, in the fridge overnight.

The next day, assess the development of your loaves. If you want to see them bigger and softer, bring them out to warm up to room temperature. Preheat the oven to 475°F (246°C), with your steam setup and baking stone, parchment-lined baking sheet or Dutch oven inside. For pseudo-historic reasons, I cut a cross on the top of each loaf, so the grandkids refer to it as "X Bread." It marks the spot! Place the loaves on your baking surface of choice (and cover with the lid if using a Dutch oven) and immediately turn the oven down to 425°F (218°C). Bake for 20 minutes, uncover the Dutch oven (if using), and continue baking for an additional 20 minutes, or until the loaves sound hollow when tapped on the bottom.

Semolina-Fennel Bread

Formula
MAKES 2 (1½-POUND) LOAVES

Water	75%	630 g
Olive oil	7%	59 g
Sourdough starter	12%	100 g
White bread flour	80%	672 g
Semolina flour	20%	168 g
Salt	3%	25 g
Whole fennel seeds	1.2%	10 g
50-50 all-purpose flour and white rice flour mix, for dusting		

Why is this bread so underappreciated? It's always the last left on the rack on a Saturday afternoon, like a scrawny kid at a pick-up baseball game. More than once, I have heard the label misread as "Salmonella-Fennel," which does rather make her sound like Cruella de Vil's sister. Or perhaps it's the American aversion to the herb itself, suspiciously strong—"Isn't it the same as licorice?"

Personally, I've always loved this one, and not just because of its underdog status. I like the sharp crust, the golden durum crumb, and yes, the little seeds that bite back just a bit. It is an amiable dough, though, and hard to ruin. It bakes well even at low temps, with dependable oven spring and nice ears (the opening of scored cuts on the crust). Nothing goes better with blueberry jam, or works as well for wiping a plate of tomato sauce clean.

Semolina is a coarse grind of flour made from hard durum wheat, mostly used in couscous and pasta. We use it here, but also in the Olive-Semolina Fougasse, in crackers, and in pasta. It is high in protein and gives a strong elasticity to any dough, as well as a beautiful golden hue. I consider it a necessary part of any baker's toolbox.

Instructions

Combine the water, olive oil, starter, and flours in a mixer fitted with the paddle attachment. Mix on first speed for 4 minutes. Add more water as necessary to keep the dough well incorporated, then cover and give it a 20-minute rest.

Replace the paddle attachment with the dough hook; oil the hook if necessary (see How to Mix with a Stand Mixer, page 48). Uncover the dough and mix on second speed for 4 minutes, adding the salt halfway through. Trickle in enough water to keep it slack. Add the fennel seeds in the final minute, along with a little more water.

Transfer the dough to a well-oiled bin and cover. Let it proof in a warm spot for 4 to 6 hours, giving the dough an envelope fold every hour or so. After it has at least doubled in volume, transfer it to a floured work surface and wipe off any excess olive oil with a paper towel. Divide the dough in half and put the loaves in their proofing baskets. Dust well, then put the baskets, covered, in the refrigerator for the night.

The next day, check on the development of the dough: If it's bulging already, leave it in the fridge until the oven's hot. If not, bring it out for a while before preheating. Set the oven to 475°F (246°C) with your steam setup and baking stone, parchment-lined baking sheet, or Dutch oven inside. Grab your lame for scoring. I carve a "baseball stitch" on this loaf: two arcs, almost touching, as if to tell this one that I want it on my team. Place the loaves on your baking surface of choice (and cover with the lid if using a Dutch oven). Immediately turn the oven down to 425°F (218°C) and bake for 40 minutes (if using a Dutch oven, 20 minutes covered and an additional 20 minutes uncovered), or until the loaves sound hollow when tapped on the bottom.

Hat Trick Bread

Formula
MAKES 2 (1½-POUND) LOAVES

Water	75%	630 g
Sourdough starter	12%	100 g
Bolted bread flour	100%	840 g
Salt	3%	25 g
Malted barley seeds	4.2%	35 g
Thick-cut oats, for dusting		

When I was growing up in suburban Montreal in the 1960s and '70s, hockey was our religion, and one of our greatest demigods was the captain of the Canadiens, Jean Béliveau. He was remarkable, as an athlete and as a man. When he died in 2014, his funeral was viewed by the entire nation on TV, and his comrade Yvan Cournoyer teared up as he lamented the loss of *mon capitaine*. This loaf was designed and named in honor of Jean Béliveau. In hockey, soccer, and cricket, where it seems to have originated, a hat trick is the scoring of three goals by one player in one game. In *this* game, the three goals are wheat, oats, and barley. The player is you!

Of course, you don't need to be a Habs fan to like this bread. Just like that Hall of Famer and Stanley Cup winner, it is elegant and full of good taste. Healthy. Well-rounded. A winner. The base is made of bolted wheat flour, with whole malted barley seeds in the crumb and oat flakes that toast on the crust. The malted barley comes from our local mill and malt house, Ground Up Grain, which also mills and bolts the flour for this recipe.

Malting is an ancient process of controlled sprouting that tricks the seeds into germinating (after a long soak), which converts starches into sugar. Before that process goes too far, the seeds are kiln-dried and made stable again, but now they are much sweeter. Generally used for brewing, malted barley has a flavor reminiscent of malted milk. Though barley bread is made in regions where wheat won't grow (say, much of Scotland or Tibet), it is extremely dense due to the lack of gluten. Oats also lack gluten. So, while barley flour or oat flour can be useful in pastry formulas, for bread it makes more sense to use these fine ingredients in a more distinctive and high-profile way. In malted seed form, the barley provides a chewy texture to the crumb. As flakes, the oats speckle the crust and give it more crunch. Plus, we source the oats from Quebec, so that's a lap around the rink right there.

Bolted bread flour is something in between white and whole wheat: not quite as elastic as the former but more extensible than the latter. So, hydrate accordingly. If your flour is freshly milled, you will notice a bump in both the enzymatic activity and the flavor. You will have a somewhat narrower window of optimal hydration and development—that is, less wiggle room in terms of both shaping and baking. When you hit it right, the loft on the wheat will astound while the other two grains skate circles around you.

Instructions

Combine the water, starter, and flour in a mixer fitted with the paddle attachment. Mix on first speed for 4 minutes, adding water if necessary. Cover and let it rest for 20 minutes.

Replace the paddle attachment with the dough hook; oil the hook if necessary (see How to Mix with a Stand Mixer, page 48). Uncover the dough and mix on second speed for 4 minutes, adding the salt halfway through and the malted barley seeds with 1 minute left. As ever, adjust the hydration, adding water sparingly to allow for a slack dough without drowning it.

Transfer to an oiled container, cover, and bulk proof for 5 to 6 hours in a warm place. Give the dough 2 or 3 envelope folds during this time to develop the gluten strands. After the dough has at least doubled in size and feels full of air but not yet sloppy and sticky, divide the dough in half and shape each into a tight round. Don't overdust your hands or the table or the dough, because you want the oats to stick to the dough when you're done. Dunk each finished round into a bowl of thick-cut oats and place gently into a basket. The oats will keep it from sticking to the sides. Cover and place in the fridge for the night.

In the morning, check on the development of your loaves. If they are bulging, then leave them where they are. If they seem a little underdeveloped, put them on the counter and give them floor time while they come up to room temperature. Preheat the oven to 500°F (260°C), with your steam setup and a baking stone, parchment-lined baking sheet, or Dutch oven inside. Don't bother to score, as the oats are resistant to branding and are quite pretty all on their own.

Place the loaves on your baking surface of choice (and cover with the lid if using a Dutch oven) and immediately turn the oven down to 450°F (232°C). Bake for 20 minutes, uncover the Dutch oven (if using), and continue baking for an additional 20 minutes.

Because of variations in home oven temps and in dough temps (whether it came straight from the fridge or had floor time) you will need to watch your loaf more closely than you are reading this recipe. Tap the bottom and listen for that hollow sound. Then let it cool. A lot. Go brush up on your French, or at least on your slap shot. *Bon appétit!*

Wheat Germ Theory Bread

Formula

MAKES 2 (1½-POUND) LOAVES

Water	75%	630 g
Sourdough starter	12%	100 g
White bread flour	100%	840 g
Salt	3%	25 g
Toasted wheat germ	3%	25 g
Wheat bran, for dusting		

Fred Stevens (my father) was a true classicist: a staunch proponent of germ theory and an admirer of elegant scientific models. I'd like to dedicate this loaf to him and his Edwardian ways.

Wheat seeds are remarkably stable packets of nutritional energy before they're split open and milled into flour. They're surprisingly egg-like, with a bran "shell," a starchy endosperm "white," and a "yolk" of oil and protein at their center. If kept dry and cool enough, they can be stored almost indefinitely—edible money in the bank.

This bread recipe mimics the quasi-atomic structure of wheat seeds in an old-fashioned way that my dad (who majored in chemistry at Rutherford's lab at McGill) might appreciate. It has a bran crust, a white crumb, and toasted bits of wheat germ floating around inside like little lost electrons.

It is another hybrid, somewhere in the nebulous but tasty middle of the whole wheat–white bread spectrum. Its highlights are an earthy (iron-enriched) coating and a heavenly interior with sweet, crunchy specks of goodness surprising you at random intervals.

In honor of Fred, I generally slather some orange marmalade on this one, or have it with a fried egg, which is something he ate pretty much every day for 99 years. That's a lot of theory.

Instructions

Combine the water, starter, and flour in a mixer fitted with the paddle attachment. Mix on first speed for 4 minutes, adding water if it seems too dry. Cover and allow to rest for 20 minutes.

Replace the paddle attachment with the dough hook; oil the hook if necessary (see How to Mix with a Stand Mixer, page 48). Uncover the dough and mix on second speed for 4 minutes, adding water if necessary (and scraping the sides of the bowl). Add the salt and a little water after 2 minutes and the toasted wheat germ after 3 minutes.

Transfer the dough to an oiled bin, cover, and let it sit in a warm spot for 5 to 6 hours, until it has doubled in volume, giving it an envelope fold every hour or so. Divide in half and shape into tight boules, avoiding contact with loose white flour, until they are round and sticky enough to dip into a bowl of raw wheat bran. When the rounds are well dusted, place them in proofing baskets. Let the covered, basketed boules spend the night slowly fermenting in the fridge.

The next day, check on their development: if they need a little more expansion, allow them to come to room temp. Preheat the oven to 475°F (246°C), with your steam setup and a baking stone, parchment-lined baking sheet, or Dutch oven inside. Use a new razor blade to cut through the coating of bran. I like to carve a spiral (start at the outermost part and work in) or a starburst. Place the loaves on your baking surface of choice (and cover with the lid if using a Dutch oven), and turn the oven down to 425°F (218°C) right away. Bake for 20 minutes, uncover the Dutch oven (if using), and continue baking for an additional 20 minutes, or until the loaves sound hollow when tapped on the bottom.

Eight-Grain Bread

Soaker

Discard sourdough starter	2%	17 g
Water	9%	76 g
Eight-grain mix	7%	59 g
Bran	2%	17 g
Total soaker	19.5%	164 g

Formula

MAKES 2 (1½-POUND) LOAVES

Water	75%	630 g
Sourdough starter	12%	100 g
Bolted bread flour	100%	840 g
Salt	3%	25 g
Eight-grain soaker	19.5%	164 g

50-50 all-purpose flour and white rice flour mix, for dusting

This is the alter ego to our French Bâtard, that crusty white loaf. This one is *almost* whole wheat, soft-crusted and very grainy. More German than French, from the other side of the Rhine. While it offends some purists, I do not use whole wheat flour for this formula. As detailed in In Between White and Whole Wheat, page 38, I prefer bolted bread flour, with a small portion of the bran removed. I find it provides for a much more extensible dough. We more than compensate for that "missing" fiber with our eight-grain soaker. This is a fermented soaker that not only predigests a blend of seven seeds and grains (the eighth is the wheat itself), but also a good portion of bran, which is rendered even more absorbable than usual, mixed as it is with some discarded leaven and left to stew overnight.

This is a hardy bread, but beware: it can be tempting to overhydrate and overproof it. The flour is water-thirsty, and if freshly milled, those enzymes (plus the action of the juiced-up soaker) can make the dough run away from you, particularly during its bulk proof. So, push this one, but push it gently. Score it somewhat lightly, and bake it on the hotter side.

Eight-Grain Mix

We make our own eight-grain mix with equal amounts, by weight, of the following:

Oats	Sesame seeds	Corn grits
Rye flakes	Millet	
Sunflower seeds	Flax seeds	

Soaker Instructions

To make the soaker, we combine a portion of discard starter (starter that is a little past its prime after the day's mixes are done) with fresh water, then add a portion of the eight-grain mix plus some bran. Mix it together in a slurry, cover, and leave at room temp overnight (or in the fridge if it's hot). It will expand, so use a large-enough container.

The next day, after the soaker has had its sleepover party, it's time to make the bread dough itself.

Instructions

Combine the water, starter, and flour in a mixer fitted with the paddle attachment. Mix on first speed for 4 minutes, then cover and give it a 20-minute rest.

Replace the paddle attachment with the dough hook; oil the hook if necessary (see How to Mix with a Stand Mixer, page 48). Uncover the dough and mix on second speed for 4 minutes, adding the salt halfway through, quickly followed by the soaker. The soaker will change the nature of the dough, adding liquid, texture, and possibly a little live leavening, so keep the finished mix a little on the drier side. Transfer it to an oiled bin, cover, and allow it to bulk proof in a warm spot for 4 to 6 hours, until it doubles in volume. Give it a gentle envelope fold every hour or so.

Divide the dough in half and shape gently into rounds. We dust this bread with a combination of white rice flour and wheat mids, a by-product of the milling, thereby sneaking back in even more of the fiber that was taken out by bolting. But you can use a mix of all-purpose flour and white rice flour. Put the dusted loaves in proofing baskets, cover them, and store them in the fridge for the night.

In the morning, check their development: If they're bulging nicely, leave them where they are until the oven is hot. If not, give them some floor time at room temperature to further proof. Preheat the oven to 500°F (260°C), with your steam setup and baking stone, parchment-lined baking sheet, or Dutch oven inside. Carve a simple sickle-moon cut in the top, leading curve of the round. Deliver it to the heat, say a little prayer, and turn the oven down to 450°F (232°C) right away. Bake for 20 minutes, uncover the Dutch oven (if using), and continue baking for another 20 minutes. Tap the bottom of the loaf to hear if it sounds hollow yet. If so, put it on a rack to cool. Try it with grilled cheese and tomato.

Colonel Buckwheat Bread

Bran Starter

Sourdough starter	500 g
Water	500 g
Bran	250 g

Formula
MAKES 2 (1½-POUND) LOAVES

Water	75%	630 g
Bran starter	12%	100 g
White flour	100%	840 g
Salt	3%	25 g
Toasted buckwheat groats	5%	42 g
Coarse cornmeal, for coating		

The name of this loaf is a dad-joke pun that all too often misses the mark: many (or most?) customers quickly scan the label and ask for the "Colonial" Buckwheat. It's meant to be a *corn* joke, as in *kernel*, but nevertheless. . . . Colonel Buckwheat is a real character, covered in coarse polenta, very crusty, but with a soft interior punctuated by toasted buckwheat groats. It's a triple combination (counting the white wheat flour), with the dark bitter tones of the buckwheat playing against the bright sweetness and texture of the corn.

It turns out that buckwheat is not a wheat at all, but rather what is ungenerously referred to as a pseudocereal. Though I've always associated it with Breton crêpes, dark honey, and soba noodles, its origins are distinctly Chinese. It flowed to Europe in separate waves—one northward through Russian peasant cuisine (kasha, blintzes, etc.) and another to France and Italy through the Middle East, where it became known as sarrazin, as in Saracen or Muslim.

As buckwheat floated to the New World on sailing ships, corn from the Americas flowed in the opposite direction. High and low palates alike adapted to corn quickly and enthusiastically. In our late-period iteration, wheat flour binds it all together and provides the medium for a conversation between the two.

So, the Colonel (or "Uncle Bucky," as one customer calls it) is a truly cosmopolitan fellow, with Mayan eyes, a Silk Road smile, and a Levantine soul. He belongs to no army but his own; his honorific is self-appointed. Perhaps he is a little colonial after all, but in a South American, Bolivarian way. This is a bread one might find inside a Gabriel García Márquez story.

And that means there's a dark secret, which in this case is an all-bran starter! What is that? Just as it sounds, it is a natural leaven using bran instead of white, rye, or whole wheat flour. It is an excellent way to reintegrate bran into a white-flour dough while allowing for extra fermentation of the bran, which in turn makes it more digestible and less disruptive to the gluten strands. Though the starter appears unimpressive, even when ripe (there's no viscosity or bubbling at all), it can easily push any dough at the same rate as regular starter, plus it adds a particular branny must to the resultant bread. It's good for muffins, too.

The Colonel is a gregarious guy. He gets along with anyone—butter, jam, olive oil, pasta sauce, anything! Be sure to use a sharp knife on this loaf, as the corn crust will repel a blunt one. As for that crust, a very coarse cornmeal seems to work best for us, though it's often harder to source, rarer than the fine cornmeal products out there. We use Bob's Red Mill polenta.

Starter Instructions

Refresh your regular starter in the usual way, with 4 parts white flour (by weight) and 1 part rye, plus an equal amount (again by weight) of filtered or at least not overly chlorinated water, all at room temperature. Leave covered at room temperature for 4 to 6 hours.

To make the bran starter, we simply take a portion of regular starter, add water and bran, then leave it overnight at room temperature. One part starter to one part water and a half part bran. Bran is dry, so mix in enough water to make it slightly less than stiff (but not sloppy). There is an element of faith here, at least for the first time, but trust me: it works! You'll need reasonably fresh bran, not something that's been lying around your pantry for years.

As mentioned, the bran starter will not have the same bubbly personality as the regular stuff. Have no fear and carry on!

Instructions

Combine the water, bran starter, and flour in a mixer fitted with the paddle attachment. Mix on first speed for 4 minutes. If the flour is particularly strong and seems dry, trickle in a little more water. Then cover and allow to rest for 20 minutes.

Replace the paddle attachment with the dough hook; oil the hook if necessary (see How to Mix with a Stand Mixer, page 48). Uncover the dough and mix on second speed for 4 more minutes, adding the salt halfway through and the buckwheat groats with 1 minute to go. As always, add water to keep the dough somewhat slack, but don't overdo it: be circumspect.

Place in an oiled bin, cover, and bulk proof in a warm place for 4 to 6 hours, until the dough has at least doubled in volume. Give it a couple of envelope folds during this time to help develop the gluten structure. When the dough is ripe—tacky but not yet sticky—divide it in half and shape the dough into rounds.

Dump each round into a bowl of coarse cornmeal and coat the entire surface. Drop these yellow babies into baskets, cover, and put them in the fridge (covered!) overnight.

The next day, take a look at your cold, sleepy dough and judge how ready-to-bake it is: Puffed up, or still shy? If shy, leave it on the counter for an hour or two. Meanwhile, preheat the oven to 475°F (246°C), with your steam setup and baking stone, parchment-lined baking sheet, or Dutch oven inside.

When you and the oven and the dough are all ready to bake, arm yourself with that miniature sword (the razor blade on a stick) and slash away. Deeper if it can take it (and at an angle) or more lightly if it seems the round is going to deflate. Introduce it to the heat, close the door, and immediately turn the oven down to 425°F (232°C). Bake for 20 minutes, uncover the Dutch oven (if using), and continue baking for another 20 minutes. If the color looks good, tap it on the bottom and see if there's a hollow sound: that's the sure sign of a baked loaf. Put it on a cooling rack and really let it cool. No nibbling until it is *actually* cool. Might take hours. Read that Márquez book.

Cornmeal Dance Bread

Soaker

Coarse cornmeal (polenta-grade)	13%	109 g
A good locally brewed ale	6%	50 g
(about 1 pint per kilo of cornmeal)		
Water	1%	8 g

Formula
MAKES 2 (1½-POUND) LOAVES

Water	75%	630 g
Sourdough starter	12%	100 g
Soaker	20%	168 g
Strong bread flour	62%	521 g
Bolted rye flour	25%	210 g
Salt	3%	25 g
Sunflower seeds, for coating		

Without music, I'd be unable to bake bread. It moves my limbs and keeps my mind bent on mixing rhyme. William Parker, formidable bassist, is one of my unwitting collaborators here: my Cornmeal Dance Bread is named for a song of his, recorded with the Raining on the Moon ensemble. In it, Leena Conquest sings: "High Sound, High Silence . . . all the world's a dream to me, except your sweet soul. The mountain is dancing." William's been in the bakery a few times, and on the rare occasion that I get to see a show of his, he always says, "Hey, Bread Guy!"

The Loaves

Aside from jazz poetry, this loaf draws inspiration from a couple of early Commonwealth recipes: Ryaninjun and Anadama. The first is a contraction of "Rye-and-Indian" since it was a combination of both native corn and transplanted rye (which grew better than wheat due to its resistance to blast fungus). John Winthrop Jr., son of the Massachusetts Bay Colony's first governor, wrote in 1665: "There is . . . very good Bread made of Indian corn by mixing half, or a thirde parte, more or less of Ry, or Wheat-Meale or Flower amongst it, and then they make it up into Loaves, adding Leaven or yeast to it to make it Rise."* Often, that yeast was found in beer barm (hence the "dancing" cornmeal here, soaked in ale overnight). Ryaninjun was usually wrapped in oak or cabbage leaves to bake.

Anadama's origins are said to be in Rockport, Massachusetts, where a "hangry" fisherman had to mix his own bread dough with leftover cornmeal and molasses, all the while cursing his wife, Anna. Perhaps. Since molasses came from slave plantations in Jamaica, we'll just skip that ingredient. Think of this bread as Anna's revenge.

We use locally milled rye flour from Ground Up Grain, as well as the People's Pint Farmer Brown Ale (from Greenfield, Massachusetts), with hops grown by our friends at Four Star Farms in nearby Northfield. This loaf also gets a distinctive and earthy crust with a coating of sunflower seeds.

---

* Keith Stavely and Kathleen Fitzgerald, *America's Founding Food: The Story of New England Cooking* (Chapel Hill, NC: University of North Carolina Press, 2004), 24–25.

Soaker Instructions

Combine the cornmeal, beer, and water in a container. Cover and let it sit at room temperature overnight.

Instructions

Combine the water, starter, soaker, and flours in a mixer fitted with the paddle attachment. Mix on first speed for 4 minutes. Cover and let it rest for 20 minutes.

Replace the paddle attachment with the dough hook; oil the hook if necessary (see How to Mix with a Stand Mixer, page 48). Uncover the dough and mix on second speed for 4 minutes, adding the salt halfway through. Add a little, but not much water; this is meant to be a heartier, denser loaf. Transfer the dough to an oiled bin, cover, and leave it in a warm place. Over the course of 4 to 6 hours, as it doubles in volume, give it 3 or 4 envelope folds.

When it has proofed sufficiently, divide it in half, delicately shape it into rounds, and coat (while sticky) with sunflower seeds. Put the rounds in their proofing baskets and leave in the fridge, covered up, overnight.

In the morning, check on its development. It won't be as proud as a white loaf, but it should have some bulge. If not, give it some floor time at room temperature. Meanwhile, preheat the oven to 500°F (260°C), with your steam setup and a baking stone, parchment-lined baking sheet, or Dutch oven inside. Due to the sunnies, we won't score this loaf, but it can often burst in a beautiful way. This one likes a hot oven, though, so check your back-up thermometer and leave the setting on high for at least 40 minutes. If it's in a Dutch oven, remove the lid halfway through.

After cooling, try it with some dried fish (such as alewife: an anadromous herring that migrates back to its river of origin) or some strong English cheese. Plus, more ale, of course!

Country Bread

Formula
MAKES 2 (1½-POUND) LOAVES

Water	75%	630 g
Sourdough starter	12%	100 g
White flour (all-purpose or bread)	80%	672 g
Rye flour (whole or bolted)	20%	168 g
Toasted rye flakes	12%	100 g
Stout or beer, for soaking		
Salt	3%	25 g
Sesame seeds, for coating		

White flour is refined, and it was long reserved for "refined" folk: rich people who could afford it, while the poorer could not. Of course, historically, rich folks also suffered disproportionately from gout and certain kinds of decadent obesity. In pre-Revolutionary France, aristocrats even put white flour in their wigs as peasants nearby were starving. It was not a good look. Guillotines ensued.

Rye flour was cheaper and more available than wheat, so "country" folk would bulk up their *pain de campagne* with it. So, our Country Bread has a fairly traditional base of white and rye flours in a 4-to-1 ratio, but with some unusual flavors and textures in both the crust and crumb. To augment the "rye-ness" of it, we put rye flakes in the dough after they've soaked awhile in oatmeal stout. Rye flakes look and act a little like cut oats, and they can be used in lots of creative ways. Beer and bread go way back, of course, to the days when beer was referred to as "liquid bread," and many bakers used brewer's barm as a leavening agent. While the alcohol is baked off in the oven, a delicious frothiness remains. We use a local favorite: the People's Pint Oatmeal Stout. I suggest that you shop (and drink!) locally and responsibly, and soak your rye flakes in whatever seems best. (We've got a vibrant population of hops growing around the front door of the bakery, in a seasonal battle with the porcelain berry vines. One customer made a peanut butter stout from them, and though it was by far the most local beer I've ever consumed, I decided it might not work well in this bread.)

The other twist is the sesame crust. As outlined in Starters, Dusters, and Offbeat Inputs on page 42, many kinds of seeds will work well to keep the dough from sticking to the baskets overnight and to add another dimension to the outer layer of a loaf. Sesame seeds are plentiful, cheap, and yummy.

There's something very grounded and strong about a bread with three main elements, a magic triangle of sorts. Wheat, rye, and sesame. Like the Hat Trick or Colonel Buckwheat loaves, the corners all face each other equally and bounce around the tongue and teeth. Maybe the Masons know about that, and perhaps that's why I score these loaves with the Greek symbol for pi: there's an infinite depth to this bread. A country pi. A big seller on March 14, when the math majors at Smith College come by. This is one of my favorite loaves, and I eat a whole one every week. Good with ripe avocado or almond butter. Sweet notes with a little tang.

Instructions

Combine the water, starter, and flours in a mixer fitted with the paddle attachment. Mix on first speed for 4 minutes, adding water if necessary. Rye flour tends to be thirsty and can make a dough fairly dry. Depending on the grind of the rye flour used, hydration will vary. Keep a pitcher of water handy. A strange thing about rye dough of any kind is that it can start to bind up and actually *repel* water if allowed to get too dry. Even a 20-minute autolyse won't allow it to relax. Something in the pentosans, the five-carbon sugars that predominate in rye flour, allows it to absorb water quickly, but only to a point. Try to stay ahead of that clay-like curve without overhydrating. Call it surfing in the mixing bowl.

Cover and let the dough rest for 20 minutes. Meanwhile, soak your toasted rye flakes in just enough stout to cover the flakes. A little extra foam won't hurt.

Replace the paddle attachment with the dough hook; oil the hook if necessary (see How to Mix with a Stand Mixer, page 48). Uncover the dough and mix on second speed for 4 minutes, adding the salt and a little water halfway through. At 3 minutes, add the rye flakes and any unabsorbed beer. Keep this dough on the wet, sloppy side. It rains a lot in the country, which is OK! Things get muddy, but that's how they'll sprout and grow.

Place the dough in an oiled bin, cover, and give it at least 4 to 6 hours to double in size, with 2 or 3 envelope folds along the way. When it has doubled and has sufficient structure, divide it in half. Form the dough into rounds, taking care to keep it sticky and away from loose flour on the table. When each boule is wound and tight and sealed, drop it into a bowl of sesame seeds and coat it completely. Then drop it into a basket as is, with no additional flour. Store the rounds, covered, overnight in the fridge.

While the dough comes to room temperature, preheat the oven to 500°F (260°C) with your steam setup and baking stone, parchment-lined baking sheet, or Dutch oven inside. Score the loaf with a sharp lame. One of the other great things about sesame seeds is how easily you can slice through (or between) them, unlike some other specialty dusters (such as oats or a millet-flax blend) that are better left completely unscored. As mentioned, I channel my inner Euclid and practice the symbol for pi (π).

Place the loaves on your baking surface of choice and turn the heat down to 450°F (232°C) immediately. Yes, the sesame seeds will pop, alarmingly. They will also toast, deliciously. If covered, take the lid off at 20 minutes. After an additional 20 minutes, check the loaf and tap it for that hollow sound. If ready, take it out and let it cool, all the way.

Fig & Sage Bread

Formula
MAKES 2 (1½-POUND) LOAVES

Water	75%	630 g
Sourdough starter	12%	100 g
Coarse white semolina flour	20%	168 g
All-purpose white flour	80%	672 g
Salt	3%	25 g
Rubbed sage	2%	17 g
Dried figs, chopped	15%	126 g
50-50 flax seed and millet mix, for dusting		

There's a narrow slot from Wednesdays into Thursdays when I do some experimenting: a small 8-kilogram mix on one afternoon that gets baked off on the next. That's when offbeat ingredients are used, such as blue cornmeal starters, or dried fruit, or both! There have definitely been some clunkers: dried apricots, for instance, seem to turn into slimy orange mucus during fermentation. Not a pretty sight on its way out to the compost. This fig bread, though, is definitely a keeper. It has some semolina flour for chewiness and color, figs for a hint of the original garden, and sage for a taste of our bakery's herb garden: an amazing thing that our friends from Abound Design, Owen Wormser and Chris Marano, dreamed up and built. The crust is made up of flax seeds and millet. The whole thing is screaming for a schmear of hairy goat cheese and a glass of grappa.

We use durum semolina, a coarse grind of a particularly hard wheat (durum) that is yellow and often associated with pasta and couscous. It is high in protein and will give a fairly tight crumb. For bread recipes it is almost always blended with white or other flours, as in our Semolina-Fennel Bread and Olive-Semolina Fougasse. In this formula it constitutes a quarter of the flour, but it has a strong impact.

We use dried fig pieces. It is said that figs are not actually fruits, but clusters of inverted flowers, which is cool . . . until you also learn that some varieties, such as the Calimyrna, are pollinated by a special wasp that dies inside the flower. That may give vegans some pause.

Sage leaves are fun to gather out in Hungry Ghost's front garden. It is time consuming, though, and seasonal, so I use a jar of rubbed sage when I have to.

The flax-millet combination for the crust is the result of a lot of trial and error. It's crunchy and unique, and it looks pebbly.

Instructions

Combine the water, starter, and flours in a mixer fitted with the paddle attachment. Mix on first speed for 4 minutes. Add a little water if it needs helps getting incorporated, then cover and give it a 20-minute rest.

Replace the paddle attachment with the dough hook; oil the hook if necessary (see How to Mix with a Stand Mixer, page 48). Uncover the dough and mix on second speed for 4 minutes, adding the salt halfway through, followed by the sage and then the figs. Trickle in more water if the dough seems too tight, but be judicious. It's always easier to add water than to backpedal and add flour to a sloppy dough.

Put the dough in an oiled container, cover, and leave it in a warm place for 4 to 6 hours, giving it a strengthening envelope fold at 1- to 2-hour intervals.

When it has at least doubled in size, gassy but not overflowing, divide it in half and shape it into rounds. Make sure the seams close. Keep the boules sticky until they're perfectly spherical, then coat them with the mixture of flax seeds and millet in a small bowl. Place the newly shaped rounds into baskets, cover, and put in your fridge overnight.

The next morning, if the loaf is not yet bulging, give it some floor time on the counter. Preheat the oven to 475°F (246°C), with your steam setup and a baking stone, parchment-lined baking sheet, or Dutch oven inside. The coating on this loaf resists scoring, so leave it as is.

Place the loaves on your baking surface of choice (and cover with the lid if using a Dutch oven) and turn the oven down to 425°F (218°C) right away. If covered, remove the lid after 20 minutes, then continue baking for an additional 20 minutes. Tap the bottom of the loaf and if it sounds hollow, take it out and let it cool thoroughly on a rack. Stroll slowly (or ride your bike!) to your favorite grocery store and get a small wedge of goat cheese. By the time you get back, it may be ready to slice.

Raisin Bread

Formula
MAKES 2 (1½-POUND) LOAVES

Water	75%	630 g
Cornmeal starter	12%	100 g
White bread flour	80%	672 g
Coarse white semolina flour	20%	168 g
Salt	3%	25 g
Anise (fresh or seeds)	2%	17 g
Raisins	15%	126 g
50-50 cocoa powder and white rice flour mix, for dusting		

A kissing cousin to the Fig & Sage Bread, this loaf uses the same blend of semolina and white flours, plus some special herbs from our beautiful garden out front of the bakery. From mid to late summer, you can see me out there every Thursday morning, gently waving competing bees away from the anise hyssop: these long, droopy, purple flowers take some effort in separating from their stems, but the ambrosial flavor is well worth it. If you lack the luxury of fresh-picked flowers, you can easily substitute some anise seeds.

There is absolutely no sweetener in this dough besides the naturally occurring fructose of the raisins, but it does have a cocoa powder crust! I pair the finished bread with peanut butter for a long hike, or with Camembert and a pear dessert wine for a fancy picnic.

Yet another wild ingredient in this one is the cornmeal starter: a simple variation in the starter feed the night before gives a fruity, almost bubble gum–like nose to a distinct leaven. Blue cornmeal is preferred, but the standard (and less expensive) yellow works almost as well. Just use 50 percent regular wheat and rye starter to feed a 50-50 mix of water and cornmeal (by weight). If it seems a little thin, add more cornmeal to thicken. Unlike the bran starter for the Colonel Buckwheat Bread, this one will definitely bubble up and overripen into liquid, if allowed. Give it 12 hours to mature.

This is almost a kind of dessert bread, slightly denser and more cake-like thanks to the semolina flour and the raisins, of course. It is crucial that you don't hold back on the amount of dried fruit. It will seem like a lot in the mix, and then not quite enough once the loaves are baked. I would also caution you not to allow the shaped loaves to ripen overnight quite as much as other breads; it's more important to catch this one on the rising curve, rather than near the lip of the fermentation wave. Scoring this raisin with a deep crosshatch could deflate a riper loaf.

Instructions

Combine the water, starter, and flours in a mixer fitted with the paddle attachment. Mix on first speed for 4 minutes, then cover and allow it to rest for 20 minutes.

Replace the paddle attachment with the dough hook; oil the hook if necessary (see How to Mix with a Stand Mixer, page 48). Uncover the dough and mix on second speed for 4 minutes. After 2 minutes, add the salt and a little water and anise. The raisins will dry out the mix, so keep it well hydrated in advance. After 3 minutes, add the raisins, adjusting the hydration as necessary.

Transfer the dough to an oiled bin and cover. Allow it to proof in a warm spot for 4 to 6 hours, giving it an envelope fold every hour

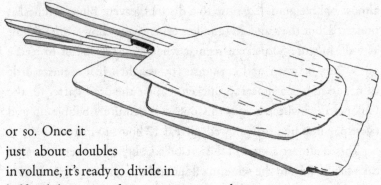

or so. Once it
just about doubles
in volume, it's ready to divide in
half and shape. Avoid any unincorporated
flour that might dry out the skin. Once it's round and sticky, drop it
in a bowl of cocoa powder and white rice flour. Coat it entirely and
drop it in a basket, as is. Let it proof, covered, overnight in the fridge.

    Check on the dough the next day. As mentioned, don't let it
balloon up quite as much as other doughs. Preheat the oven to 475°F
(246°C), with your steam setup and baking stone, parchment-lined
baking sheet, or Dutch oven inside. Score the loaf (I make a tic-tac-toe
field, minus the plays) and pop it in! Turn the oven down to 425°F
(218°C) immediately. After 20 minutes, uncover the Dutch oven (if
using) and continue baking for an additional 20 minutes. Let it cool
and "X" into it with a serrated knife; your mouth will be the "O."

Rye Bread

Formula
MAKES 2 (1½-POUND) LOAVES

Water	80%	672 g
Sourdough starter	12%	100 g
Bolted rye flour	50%	420 g
Strong white bread flour	50%	420 g
Salt	3%	25 g
Caraway seeds	2%	17 g

50-50 rye flour and white rice flour mix, for dusting
Kalonji seeds, for sprinkling

Rye is another one of those less-glutenous cousins of wheat, in the vein of spelt and khorasan, but with more nostalgic associations, at least for many Americans of Central European ancestry. Pumpernickel, Jewish rye, Finnish rye, Swedish rye … *rye all the distinctions?* Because traditions deepen practice, and practice makes do with what's at hand. Many people like to sweeten up their rye dough, while others darken it with coffee or molasses. Some like it super dense and sliced extra thin. Almost all of them, by necessity, require loaf pans because of the lack of gluten. Instead, rye has pentosans that swell with water and thus lighten up the crumb.

Rye has flavors and beautiful, bitter depths that wheat just lacks. It grows in wetter, colder, snowier climates and in poorer soils. It fixes nitrogen in the soil. Its higher amylase enzyme content speeds up fermentation. At Hungry Ghost, we use rye flour (25 percent) in our twice-daily starter feeds, and we make rye crackers (50 percent) with dried onion, our Cornmeal Dance Bread and Country Bread each with 20 percent bolted rye flour, and this Rye Bread with a 50-50 blend.

Our Rye Bread is a happy compromise with strong white bread flour that makes a slightly denser but still proud and airy loaf. It's a hearth loaf, with enough structure to avoid the dreaded (by me) loaf pan. We put caraway seeds inside and kalonji (aka nigella or charnushka) seeds on top. Some of our older German customers swear by it, and some have even taken loaves back to Germany when visiting! Sad to remember, but one of my favorite customers ever, Friedl, who would insist on a new poem with every purchase, actually passed away peacefully in her car, parked right across the street, after getting her last loaf of this rye. I think of her every time I bake it.

Instructions

Combine the water, starter, and flours in a mixer fitted with the paddle attachment. Mix on first speed for 4 minutes. Be ready with a pitcher of water handy, because your flour may be quite thirsty and require more hydration quickly. When rye flour binds up, it actually begins to repel water and needs to rest a while before effective hydration can begin again. Cover and let rest for 20 minutes.

Replace the paddle attachment with the dough hook; oil the hook if necessary (see How to Mix with a Stand Mixer, page 48). Uncover the dough and mix on second speed for 2 minutes. Add water if necessary to keep it wet enough to accept the salt and caraway seeds, then continue for a final 2 minutes. Adding water is fine; you may go as high as 90 percent hydration. It's not lack of structure that will flatten this loaf (as in a spelt), but overfermentation.

Transfer the dough to an oiled bin, cover, and set in a warm place to bulk proof. Give it occasional envelope folds, though these aren't as necessary or useful as they are with the other types of dough. Keep an eye on it, as it will ripen faster than an all-wheat dough. Somewhere between 4 and 5 hours, when the dough has expanded by 1½ to 2 times the original volume, it will be time to shape it.

This is a sticky dough, more like a clay. It won't stretch like a full wheat dough, but it will form a ball. Divide the dough in half and shape into rounds, making sure to pinch the seam tightly closed. We basket these differently than other rounds, using linen basket liners (old ripped-up couches, in fact) that get dusted lightly. Because I load them into a very hot oven as quickly as possible the next morning, I take the time to spray the tops with water and spread kalonji seeds on top before covering and putting them away in the fridge for the night.

The next morning (or at least 12 hours later), check the dough development: Are they bulging out of the baskets? If not, give them an hour or more of warm room time. Preheat the oven to 500°F (260°C), with your steam setup and baking stone, parchment-lined baking sheet, or Dutch oven inside. I slash these loaves with a cursive T. Brush off what excess dusting flour you can as you ease the dough onto your baking surface.

Turn the oven down to 450°F (232°C) right away. Bake for 40 minutes, uncovering the Dutch oven (if using) halfway through. Check the color and the sound of the loaf. If it's well darkened and hollow sounding, you're good to go!

Cooling the rye loaves is even more important than with wheat. They need at least a full day to cure and settle into themselves. In some regions of Germany, I'm told, you have to let them sit for a week or more! Thanks to those pentosans, rye will last much longer than wheat bread. I take a number of them on biking trips and they last two weeks or more. But I need to remember to bring my folding serrated knife.

Khorasan Bread with Pumpkin Seeds

Formula
MAKES 2 (1½-POUND) LOAVES

Pumpkin seeds	3%	25 g
Water	83%	697 g
Spelt or regular sourdough starter	12%	100 g
White khorasan flour	100%	840 g
Salt	3%	25 g

50-50 all-purpose flour and white rice flour mix, for dusting

Khorasan is an older form of wheat, a hard durum, that originated in what is now the province of Khorasan in northeastern Iran. It grows only in dry, hot zones. For some, its claims to fame are that it is ancient and lower in gluten, but I bake with it because it tastes good and makes a very different kind of loaf. It has a nutty, buttery taste, with a thinner crust and a slightly denser crumb. It has the highest hydration of any of our loaves, starting at 83 percent and adjusted upwards from there. In part thanks to this, I find that I can bake it toward the front of my Saturday morning bakes, in a hotter zone, unlike some of the other white breads. It generally won't have a lot of oven spring, but it does caramelize quite nicely.

You can certainly use your regular wheat and rye starter for this formula, but we dip into our (less-used) spelt starter for this loaf (and for the Spelt Bread, naturally). For our customers who want to avoid wheat as much as possible, it's certainly more consistent. If you want to establish a spelt starter, simply take some regular wheat and rye starter and feed it with water and spelt flour in a separate container. You could do this multiple times to diminish the residual wheat content, or simply wait one cycle. If you're using white spelt, up the ratio of flour to more (per weight) than the water, as it needs to be even stiffer than the regular stuff.

This is an offbeat but very rewarding loaf, punctuated with green pumpkin seeds throughout. I recommend using organic pumpkin seeds if possible. If not, rinse them before you soak them. A softer loaf, this one lasts for days and toasts well even a week later. It goes equally well with baba ghanoush or a fig compote.

Instructions

Get your pumpkin seeds soaking before you begin mixing. Cover them in warm water and let them sit.

Meanwhile, combine the water, starter, and flour in a mixer fitted with the paddle attachment. Mix on first speed for 4 minutes; if it still looks dry after the first 2 minutes, add more water. As noted, this is easily the "thirstiest" dough we work with. Don't be afraid to water this plant.

Cover and give it the customary 20-minute rest. Meanwhile, drain your pumpkin seeds and get your salt ready, as well as a pitcher with even more additional water, in the likelihood that this near-desert grain is still parched.

Replace the paddle attachment with the dough hook; oil the hook if necessary (see How to Mix with a Stand Mixer, page 48). Uncover the dough and mix on second speed for 4 minutes, adding the salt and a little water after 2 minutes and the pumpkin seeds after 3 minutes. It will seem sloppier than the other bread doughs. That's OK!

Transfer the dough to an oiled bin, cover, and let it proof for 4 to 5 hours at room temperature. Give it an envelope fold every hour or so, if possible. When it has just about doubled in volume, divide the dough in half and shape into rounds. It will feel loose and a little challenging to shape. Don't overdo it; simply coax it quickly (with minimum contact) into a sphere, then scoop it up with your dough knife, sprinkle very generously with your dusting mixture, and dump into a basket. Add more duster around the edges of the loaf where it will expand and lean into the walls of the basket. This is a sticky dough, so you want to do all that you can to ensure that it will emerge unscathed when it's finally time to bake. Cover the shaped loaves and put them in the fridge overnight.

Preheat the oven to 475°F (246°C) with your steam setup and baking stone, parchment-lined baking sheet, or Dutch oven inside. Take the dough out to assess its development. If needed, give it some floor time to come up to room temperature and expand a little further. I score this loaf somewhat shallowly, in the shape of the letter K, right before loading it in. Turn the oven down to 425°F (218°C) immediately. After 20 minutes, uncover the Dutch oven (if using) and bake for an additional 20 minutes, then tap the loaf to see if it's ready. If it sounds hollow, then the gelatinization is complete, and you have yourself an edible piece of geopolitical archeobotany to savor!

Spelt Bread

Formula
MAKES 2 (1½-POUND) LOAVES

Water	70%	588 g
Spelt starter	12%	100 g
White spelt flour	65%	546 g
Bolted spelt flour	35%	294 g
Salt	3%	25 g
Chamomile flowers	10%	84 g
Quinoa flakes, for coating		

The central fact about spelt grain is that it is an older cousin of wheat that goes a little further back in the *Triticum* generations, so it still has a hull that covers the seed. Furthermore, the proteins in that seed are not glutens as much as they are gliadins: this is both good news and bad news. As Cheryl always says, "Everything has its antithesis." The good news is that spelt can often grow in warmer, wetter climates and is easier to digest. The bad news is that you have to dehull the seeds, and also that making bread from spelt flour is very tricky.

The great thing about wheat gluten is that it has both extensibility and elasticity: it will spread out, and it will spring back. Gliadins lack elasticity, so they will only spread out. That's why most spelt breads are made in a loaf pan. Our Maine-based sister bakery, Tinder Hearth, uses a spelt dough that pours like batter—and their bread is delicious! But since I don't have any use (or room) for loaf pans, and I like a good challenge, we make a 100 percent spelt hearth bread. After years of experimenting,

I use a blend of bolted and white spelt flours (whole spelt is great for some pastries, but a little too heavy for bread, in my opinion).

The crucial thing is to get the hydration right, since spelt dough has much narrower parameters for error. It is, unfortunately, easy to have either too much or too little water in this dough. Too much turns the loaf into a pancake; too little turns it into a brick. But as Goldilocks says, just right is my kind of thing!

To make a spelt starter, take some regular wheat and rye starter and feed it with water and spelt flour in a separate container. You could do this multiple times to diminish the residual wheat content, or simply wait one cycle. If you're using white spelt, up the ratio of flour to more (per weight) than the water, as it needs to be even stiffer than the regular stuff.

Our Spelt Bread has a dedicated bunch of followers who line up every Saturday morning. This dough is challenging to mix, hard to shape, finicky in proofing, and requires neither too hot nor too cold a bake, but it can calm a whole family of bears when done properly, especially with a dose of chamomile flowers inside. Plus, unlike wheat breads, it can be safely sliced up when still warm. I enjoy it best with a bit of orange marmalade.

Instructions

Combine the water, starter, and flours in a mixer fitted with the paddle attachment. Mix on first speed for 4 minutes, or until well incorporated. Cover and let it rest for 20 minutes.

Replace the paddle attachment with the dough hook; oil the hook if necessary (see How to Mix with a Stand Mixer, page 48). Uncover the dough and mix on second speed for 2 minutes. Trickle in enough water to get the dough to spin out into a star shape (on a countertop mixer, it may look more like a figure skater, with arms and a leg flung out). Add the salt and chamomile flowers and a little water. Mix for another 2 minutes;

try to maintain the star shape with a bit of water, if necessary. It should be a little looser than most bread dough, more slippery, but not a batter.

You'll notice the extensibility as you coax it into an oiled rectangular container. Give it an envelope fold right away, then cover and fold again each hour or so (more often than a regular bread dough). It wants the attention. After 6 to 7 hours in a warm spot, if it has doubled in volume and feels pillowy, then it is time to divide the dough in half and shape it.

Some of our veteran shapers just walk away when it's spelt time. They dislike the texture and finickiness. Others gravitate to it. It requires a firmer, quicker hand. Spelt dough is both odd and forgiving: it seems to handle sutures and appears self-healing at times after it begins to tear. A characteristic of its idiosyncratic proteins, no doubt.

To compensate for its stickiness, we coat the dough in quinoa flakes. That gives it a sweet crust and keeps it from adhering to the sides of the basket. We also tend to put these loaves seam-up in the baskets, because the flakes have a way of keeping them from being all the way sealed. Cover the loaves and store overnight in the refrigerator.

The next day, if the loaves are bulging to the tops of the baskets, leave them in until the oven is hot. If not, take them out for an hour or so. Preheat the oven to 475°F (246°C) with your steam setup and baking stone, parchment-lined baking sheet, or Dutch oven inside. Flip your loaves over, gently, and brush off any excess quinoa flakes. Cut a decisive slash ¼ inch into the loaf in a sickle-moon swath. Coax the whole thing onto your baking surface, and cover if using a Dutch oven. Turn the heat down immediately to 425°F (218°C). Uncover the Dutch oven (if using) after 20 minutes, then continue baking for another 20 minutes. Check the loaf; it should be dark brown (thanks to the quinoa) and hollow-sounding. Let it cool—though it can handle warm cutting for some odd reason!

Cranberry Holiday Bread

Formula		
MAKES 2 (1½-POUND) LOAVES		
Egg	1 large egg (1 egg per kilo of flour)	
Honey	4.5%	38 g
Olive oil	3%	25 g
Water	64%	538 g
Sourdough starter	12%	100 g
Strong white bread flour (14% protein)	80%	672 g
Semolina flour	20%	168 g
Salt	3%	25 g
Fresh or thawed frozen whole cranberries	18%	151 g
Poppy seeds, for coating		

This loaf has seen some different iterations, not a few of which had wild rice in them, but that often made the crumb sticky and starchy. We make it only for Thanksgiving and Christmas, two of our busiest times. We are honored to be part of so many family feasts, though I have to say that we're pretty exhausted by the time our own holiday meal finally comes around! We're getting a little long in the tooth for these all-nighters, blasting out *The Last Waltz* and every Arlo Guthrie album ever made in order to keep up the momentum. One strategic adaption was to make this an enriched dough that allows it to be easily shaped and proofed on the Big Day (after an overnight bulk ferment), then baked on a lower-temperature tandem bake (skip that last fire, go home early!). It also gives it a longer shelf life.

Imagine, then, a fusion of Narragansett and Italian ingredients, a New World panettone with cranberries and semolina flour. Italian not for Columbus, so much, as perhaps Giovanni Caboto (John Cabot), who really did make it to Massachusetts, as far as we can tell, back at the end of the fifteenth century. I'd like to think they all would have given thanks for a bread like this. This is a golden challah dough with big chunks of fruit jammed in there and celebratory poppy seeds coating the whole crust. It's perfect for cold turkey sandwiches many days later.

Instructions

Beat the egg, then whisk in the honey and olive oil. Combine the water, starter, and flours in a mixer fitted with the paddle attachment, then add the egg mixture. Mix on low speed for 4 minutes, enough to incorporate well. Cover and let rest for 20 minutes.

Replace the paddle attachment with the dough hook; oil the hook if necessary (see How to Mix with a Stand Mixer, page 48). Uncover the dough and add the salt and a little water right away on first speed, then gear up to second speed for 4 minutes. Pour in the cranberries halfway through. Yes, it will seem like a lot of fruit, and some of those little guys will escape and roll away later, when you are shaping the dough, but remember that if you are not overstuffing it, you will be underwhelmed. As an infamous celebrity once asked me on the phone, "Where did all the olives go?" (And I replied that they were like wealth, not evenly distributed throughout the population.)

Transfer the dough to an oiled bin, cover, and bulk proof overnight in the fridge—or outside, if it's between 32 and 42°F (0–5.5°C).

Divide the dough in half and shape into rounds, dusting with poppy seeds to coat them and give them even more panache. Put them in baskets, covered, for 4 to 5 hours.

Preheat the oven to 375°F (190.5°C), with your steam setup and a baking stone, parchment-lined baking sheet, or Dutch oven inside. When the loaves have filled out their baskets, maneuver them gently onto your baking surface. Score the dough with the innards of a peace sign (the loaf itself is the outer circle); we all could use some more peace, around the table and around this weary world. I'll give aspirational thanks for that. Bake for 20 minutes, uncover the Dutch oven (if using), and continue baking for an additional 20 minutes, until hollow-sounding when tapped.

Challah

Formula

MAKES 2 (1½-POUND) LOAVES

Egg	1 large egg (1 egg per kilo of flour)	
Honey	4.5%	38 g
Olive oil	3%	25 g
Water	64%	538 g
Sourdough starter	12%	100 g
Strong white bread flour (14% protein)	100%	840 g
Salt	3%	25 g

White rice flour, for dusting

Egg wash (1 egg whisked with a splash of cold water)

Sesame seeds and/or poppy seeds (optional)

Challah is definitely one of those breads imbued with surplus symbolism and fraught with meaning. It is, primarily, the center of the Shabbos meal on Friday nights for observant Ashkenazi Jews. It is braided, an endless loop, like the cycles of time that it marks. It is a bulging, bountiful offering, in the spirit of thanks with which it is blessed and raised above the head and then torn apart by a *minyan* of eaters. It is an honor to shape and bake these edible prayers for our community.

At the same time, I am conscious that there is nothing exclusively tribal about these loaves: Eastern Europeans of all stripes eat a similar (or identical!) loaf at Easter (Greek Orthodox) or any day of the week. Whenever a customer asks for Challah on a different day, I can usually guess that they are Polish.

For better or worse, I was raised in a fiercely post-tribal household. My parents were both compassionate skeptics, fundamental atheists, and empirical scientists. Yet, some search for unclassifiable meaning persists. One of my two sons and one or possibly two of my great-grandfathers were Bar Mitzvahed, so I think that makes me a semi-Semite.

So, we all project meaning outward and find it reflected back at us, even (or especially) in bread. Let us at least enrich our gestures with generosity for others. Though we can joke about the "shiksa surcharge" for those who pronounce it "TCHA-laa," we should really just bite our tongues.

Challah is an enriched dough, made with eggs, honey, olive oil, and a strong bread flour, and leavened naturally. Some will object to sourdough challah, but their evocations of the Old Country usually go back only as far as New Jersey or Long Island. I seriously doubt anyone in the shtetls had packages of Fleischmann's yeast. So, a naturally leavened challah it is, made more nutritious and longer lasting. By Sunday morning it will still be great for French toast, if there's any left over by then.

Challah dough is much stiffer than our other bread dough, with stronger flour and a lower hydration rate. Also, this is one loaf that we retard overnight in bulk and shape on the day it is baked.

This is most definitely a bread for nonvegans, as well: enriched with eggs and honey, with an additional egg wash on top, we acknowledge the contributions of the chicken-birds and bees. By their labors ye shall taste them.

Instructions

Beat the egg, then whisk in the honey and olive oil. Combine the water, starter, and flour in a mixer fitted with the paddle attachment, then add the egg mixture. Mix on low speed for 4 minutes, enough to incorporate well. Cover and let rest for 20 minutes.

Replace the paddle attachment with the dough hook; oil the hook if necessary (see How to Mix with a Stand Mixer, page 48). Uncover the dough and add the salt right away, while on first speed, then gear up to second speed for 4 minutes. Resist the temptation to add as much water as you would for the other formulas; a slight trickle to keep the mixer arm moving will suffice. Again, this dough will be much stiffer than most other breads. Place in a lightly oiled bin, cover, and give it an hour or two in a warm spot.

Braiding the dough the next day takes a little practice, in part because it's a little counterintuitive. First, divide the dough into three equal chunks, then shape each into an elongated cylinder. Try to match them in length and width. Sprinkle with flour, then line them up side by side, seam-side down. Starting at the middle, begin braiding to one end of the cylinders, but not too tightly. You are not a hair stylist, so don't pull the strands too hard. The braid needs to be snug, but not strained. Braid down to a tapered tail at the end, then seal that between your fingertips and palm and tuck it under. Finish the other half of the braid, going up toward the top, being careful to keep the same pattern going, essentially doing it backward, as if in a mirror. Seal the new end and tuck it under, too.

We used to roll raisins into some of the dough at shaping time, but this tends to distort the braid. At Rosh Hashanah, we swirl one long "snake" of dough into a rising spiral and tuck the tails in, out of sight.

Carefully place the finished braid on a couche or other cloth dusted with white rice flour, and set it in a warm spot for 4 to 5 hours. When it has bulged outward but not yet doubled in size, preheat the oven to 375°F (190.5°C), with your steam setup and a baking stone or parchment-lined baking sheet inside. Make your egg wash and, with an obstetrician's care (or with a transfer paddle), ease the loaf onto a peel. Brush lightly with the egg wash, covering the entire top, and then sprinkle with sesame seeds, poppy seeds, or both (or neither!). The egg wash will dribble and stick, so glide the whole thing into the oven as soon as possible. Bake for about 35 minutes, or until golden brown. Shabbat Shalom.

Fougasse

scything wheat
heads on this
flatbread field

It's pronounced "foo-GAS," has nothing to do with the punk band from DC (Fugazi), is a derivative of the Latin word *focus* (which means *hearth*, the focus of any home), and is the French cousin of focaccia. It's the Provençal version of flatbread, often baked at high temperatures and used to test the heat on a wood-fired oven before putting in the more delicate rounds.

Flatbreads rule, really. You find them all around the word: tortillas, chapatis, pita, and the like. They provide a plate, spoon,

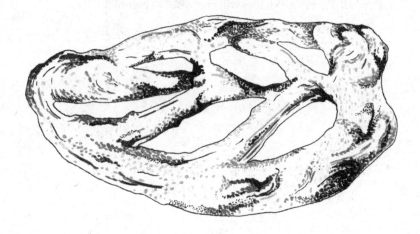

and napkin for many a meal. Ours are all wheat-based. They aren't so different from our other bread formulas, though they all have a higher hydration and are bulk-proofed overnight and then shaped right at the mouth of the oven. With their large surface area, they don't keep as long and are usually eaten, torn apart by hand, on the day of baking.

Flatbreads allow for more experimentation with flour types and extra ingredients, since the emphasis is on the crust rather than the crumb (as in a boule). One can use heavier flours and denser additions than in a round loaf. Our expanding cast of characters here includes shredded beets, coriander seeds, potatoes, garlic, dried onion, yogurt, olives, and, for the coastal types, sea water!

Aesthetically, flatbreads provide a broader "canvas" for artistically minded bakers. Some might draw elaborate motifs in the dough just before sliding them into the oven, while others sprinkle seeds, edible flowers, or strategically placed olives on the face of the loaf. Some even use a stencil, if they're not baking too many at a time. Be aware, though, that the vagaries of heat and steam will affect your creation, and not all colors and shapes will come out as planned.

Beet & Coriander Fougasse

Formula
MAKES 2 (1½-POUND) LOAVES

Clean ocean water	80%	672 g
(or plain water plus 3% salt)		
Sourdough starter	12%	100 g
Whole wheat lammas	100%	840 g
(or other heirloom variety)		
Shredded beets	30%	252 g
Coriander seeds	0.9%	8 g

This is a somewhat reverse-engineered bread, designed to take advantage of a scalding-hot oven at the start of our busy Saturdays. Made with bolted bread flour and full of wet, raw bits of shredded beet, it can withstand higher temps than any other dough I've ever worked with. Though we've experimented with carrots, yams, and fresh corn kernels, nothing compares in performance or taste to the beets. Long live the Beets!

Whole coriander seeds are a somewhat controversial ingredient here, but though we all know there are supposedly folks who are congenitally inclined to dislike the sprouted form of cilantro, these perfect little flavor bombs in the bread don't seem to taste like soap to anyone. "Weird" can indeed be very good. In order to fill the oven as quickly as possible, I scale over 20 of these at a time, set them on 8 large peels at once, and rush them into a 590°F (310°C) oven. All the while listening (every Saturday!) to Handel's *Dixit Dominus,* a scary, Goth-sounding setting of the 109th Psalm. It gets the job done.

Lammas is the name of both an ancient English harvest festival (marking the first use of the summer's new wheat) and a kind of wheat itself. The first wheat, in fact, grown in Massachusetts, imported and planted by the Pilgrims as early as 1602. We are lucky enough to have a local farmer (and amateur historian), Alan Zuchowski, who cultivates this exact variety just a few miles away from the bakery, across the Connecticut River in Hadley. It is a rather weak flour, and as such does not have enough structure for a stand-alone hearth loaf. Modern tastes often seem to require a lofty crumb while also insisting on a whole-grain base: a frustrating contradiction. Softer, more flavorful wheat such as this can be used very successfully in crackers and flatbreads.

Another left-field ingredient here is ocean water. A secret apparently known to French Navy cooks for generations is that sea water has exactly the right salinity for bread dough. Why take the salt out, only to put it all right back in again? It is a rare privilege to combine two elemental notions of home, the ocean and fresh bread, in one simple gesture. The trick is finding *clean* sea water—ask a deep-sea fisherman to fill a bucket for you, way outside the harbor. (Otherwise, plain filtered water plus your favorite salt will do just fine.)

The beets are used raw and grated up—nothing could be easier or earthier. Beets add moisture, fiber, sweetness, and a resistance to high oven temps. Whole coriander seeds keep it multidimensional, as if the Puritans made it to Plymouth via Sumatra instead.

Instructions

Combine the water, starter, and flour in a mixer fitted with the paddle attachment. Mix on first speed for 4 minutes, adding water as necessary. Don't be too timid; this is a very slack—that is to say, wet—dough. Cover and let it rest for 20 minutes.

Replace the paddle attachment with the dough hook; oil the hook if necessary (see How to Mix with a Stand Mixer, page 48). Uncover the dough and mix on second speed for 4 minutes. If you're using fresh water, add the salt after 2 minutes, along with the coriander seeds. Add the beets after 3 minutes. This will make it seem like a murder scene; after handling it, you will look like Lady Macbeth. Keep mixing for one last minute.

Transfer to an oiled bin, cover, and set in a warm spot. Fold it as best you can, a couple of times, over the course of 1½ hours. It will be sloppy and strange and staining. Cover and put it in the fridge overnight.

Preheat the oven to 530°F (276.5°C), with your steam setup and a baking stone or parchment-lined baking sheet inside. Take the dough out of the fridge and maneuver it onto a well-floured surface. Use a dough knife to divide it in half. Flour your hands well, and use the dough knife to lift the sloppy dough onto a lightly floured peel. Stretch it out into a rectangle about 1 inch thick. Flour-dust the top, lightly. Then, use the dough knife to cut two parallel lines in the rectangle, lengthwise, leaving three even sections of dough, though still connected, like a box with slits. To my mind, it evokes an African mudcloth pattern.

Jiggle the peel back and forth lightly, just to make sure the dough will slide off when you need it to. Don't dally too long—it's sticky stuff! Use a floured knife to unglue any problem spots.

Slide each dough into the oven. After 10 minutes, turn the oven down to 475°F (246°C). After another 20 minutes, check the bread: it's almost ready. The deep reds will have turned somewhat brown. Let any residual steam out (don't let it scald your face!) and give it another few minutes.

Wearing thin cotton gloves, tap the loaf on the bottom—like a newborn baby—to see if it's ready for the world. You're looking for a nice hollow sound. A dull thud means you need to send it back in for a few.

Let it cool completely. Rip it up with your hands or even slice it horizontally. The one disappointing thing about this bread is that the finished product is less lurid than the dough. The magenta tones have darkened and mellowed in the baking process. It's still beautifully flecked with beet. Perfect with a bowl of homemade hummus.

Olive-Semolina Fougasse

Formula		
MAKES 2 (1½-POUND) LOAVES		
Water	80%	672 g
Olive oil	7%	59 g
Sourdough starter	12%	100 g
Coarse semolina flour	50%	420 g
Strong white bread flour	50%	420 g
Salt	2%	17 g
Dried onion	1%	8 g
Pitted kalamata olives	12%	100 g

This is the salty-crunchy lover's dream. One customer hides it from the rest of his family behind the fridge! Better to share it, though. With semolina flour, whole pitted olives, and dried onion, baked crispy brown (don't underdo it!), it is absolutely irresistible. The only challenging parts are using the dough knife to cut a wheat head motif and sliding your sculpted dough intact into the oven. Keep the dough well hydrated, and make sure the olives are pitted.

Instructions

Combine the water, olive oil, starter, and flours in a mixer fitted with the paddle attachment. Mix on first speed for 4 minutes, then cover and allow the dough to rest for 20 minutes.

Replace the paddle attachment with the dough hook; oil the hook if necessary (see How to Mix with a Stand Mixer, page 48).

Uncover the dough and mix on second speed for 4 minutes. Add the salt halfway through, followed by the dried onion. With 1 minute left, toss in the olives. This dough does want to be on the sloppier side of things, but not soupy. Transfer to an oiled bin, cover, and give it at least an hour in a warm spot, then put it away in the fridge overnight.

Preheat the oven to 450°F (232°C), with your steam setup and a baking stone or parchment-lined baking sheet inside. Divide the dough in half, keeping the pieces together as much as possible. Stretch each chunk out on the back of your hands, pizzaiolo-style, as evenly as possible, gently repositioning your knuckles as necessary. Then turn it out gently onto a floured peel, spreading it even further to the edges without tearing or thinning out any one spot too much. I tend to keep all of our fougasses rectangular, to accommodate the different cuts that I do. For this one, we mimic the most iconic and traditional cut, that of the wheat head, most easily achieved by creating two sets of 3 diagonal cuts. However faintly, it evokes the rows of seeds on the end of a stalk, waving in the wind. Try to keep your cuts clean and decisive, using a light sprinkle of added flour if necessary. The cuts should also be as parallel and evenly spaced as possible so that no overstretched segment will bake too fast and get singed in the oven.

Jiggle it back and forth a bit on the peel so that you know it isn't sticking. When it can slide off cleanly, maneuver it to your baking stone and back off. It will bake for 30 minutes—enough time to open a bottle of wine and let it breathe while you set up to watch Marcel Pagnol's 1938 film *La Femme du Boulanger* (The Baker's Wife), which features a supremely consequential fougasse.

Potato-Thyme Fougasse

Formula
MAKES 2 (1½-POUND) LOAVES

Potatoes, cooked	57%	479 g
Garlic, minced and sautéed	1.5%	13 g
Dried thyme	1.2%	10 g
Water	80%	672 g
Olive oil	7%	59 g
Sourdough starter	12%	100 g
Strong white bread flour (12–14% protein)	100%	840 g
Salt	3.75%	32 g

The time for Potato-Thyme Fougasse is always the end of my work week, late Saturday afternoon. This is a dough that is mixed the day before, scaled out Saturday afternoon, then hurriedly and slapdashedly ladder-shaped on a peel on its way into the oven. The potatoes are cooked (boiled and mashed with garlic) two whole days before the bake. Combined with overnight fermentation and white flour and olive oil, this is a bread that can bake at almost any temperature! Whether the oven is scorching or merely tepid, only the bake time and texture vary. The real danger with this bread is biting into it while it's still hot (read: "not done baking"!). We are constantly reminding our customers to refrain from that temptation, and still we get the occasional phone call about "underbaked bread." As if!

To Canadians (*comme moi*), potatoes evoke neither Peru (their origins) nor Belgium (their apotheosis). Our potatoes come from Prince Edward Island, that (once) bridgeless birthplace of the Confederation. What a perfect metaphor for our national earthy blandness, then: a starchy but wholesome canvas to be projected onto with butter, sour cream, salsa, or curry powder. Or thyme itself. Note that the salt in this formula is at a higher percentage because of the potatoes.

This flatbread seems to have a fan base all its own, emerging out of the darkening mid-weekend skies to juggle hot paper bags right out of the door. Was there an Instagram alert? A carbohydratical pheromone wafting through the town? Who knows. It must be near 4 o'clock, because the line of devotees is forming. I try to remember to put one aside for ourselves. One quarter, sliced sideways, makes a perfect veggie burger bun.

Instructions

Let your cooked potatoes cool, then lightly mash them with the sautéed garlic and thyme; set aside.

Combine the water, olive oil, starter, and flour together in a mixer fitted with the paddle attachment. Mix on first speed for 4 minutes, then cover and allow a 20-minute rest.

Replace the paddle attachment with the dough hook; oil the hook if necessary (see How to Mix with a Stand Mixer, page 48). Uncover the dough and mix on second speed for 4 minutes, adding the salt halfway through, then the potato mixture. Sloppy but not soupy: that's our motto. Transfer to an oiled bin and cover. Give it an hour's rest in a warm spot, then put it in the fridge overnight.

Preheat the oven to 450°F (232°C), with your steam setup and a baking stone or parchment-lined baking sheet inside. Divide the dough in half, keeping the pieces as intact as possible. Stretch out the dough, evenly, on the backs of your hands. Reposition your knuckles to enlarge the surface area as consistently as possible. Turn it out onto a floured peel and keep stretching into a vaguely rectangular shape. Sprinkle a little flour on top so that the cuts you're about to make remain open. Make three stacked, parallel incisions with the full length of your dough knife, at a slight diagonal. The resultant shape, pulled a little lengthwise, should resemble a ladder. Slide that ladder gently onto the stone and bake for 35 minutes, at least. Pull it out as it gets dark brown. And let it cool!

Naan

Formula
MAKES 2 (1½-POUND) LOAVES

Water	60%	504 g
Yogurt (sheep or goat, if possible)	10%	84 g
Olive oil	7%	59 g
Sourdough starter	12%	100 g
White khorasan flour	60%	504 g
White spelt flour	40%	336 g
Salt	3%	25 g

Additional olive oil, for dipping

Za'atar blend, for generous sprinkling on top

If you had a specialized Persian oven with small stones all over the hearth, you could call this an authentic naan sangak, with its textured crust. But we'll just have to pebble the top of the dough with our fingers after we stretch it out on the peel, right before we spread some olive oil and za'atar over it. Spices travel so that we don't have to; the mind can fly far on its tongue. Za'atar is a kind of wild thyme, and also the name of a blend of herbs that includes wild thyme, marjoram, sumac, salt, and sesame seeds, though the blend will vary and often be proprietary. Our friend Suleiman Mourad brings back his Lebanese village's za'atar each year to share. Adding to the Central Asian dimension of this bread is the use of yogurt, as well as spelt and khorasan flours, both from the Fertile Crescent. *Let us break bread together.*

Instructions

Combine the water, yogurt, olive oil, starter, and flours in a mixer fitted with the paddle attachment. Mix on first speed for 4 minutes, then cover and give it a 20-minute rest.

Replace the paddle attachment with the dough hook; oil the hook if necessary (see How to Mix with a Stand Mixer, page 48). Uncover the dough and mix on second speed for an additional 4 minutes, adding the salt halfway through. Sloppy, but not soupy! Transfer to an oiled bin and cover. Give it an hour in a warm spot, then put it in the fridge overnight.

Preheat the oven to 425°F (218°C), with your steam setup and a baking stone or parchment-lined baking sheet inside. Divide the dough in half, keeping each as whole as possible. Stretch out each piece using the backs of your hands, like a pizzaiolo. Keep the dough widening as evenly as possible, so there are no bald spots (they burn!). Turn it out onto a floured peel. Push it out into a vague snowshoe shape, as the great bread ethnologist Jeffrey Alford once described it to me. Then dip your fingertips (all ten, or however many you happen to have) into olive oil and press them gently but firmly into the dough, dimpling it effectively. Sprinkle the za'atar thinly and evenly. Maneuver the dressed-up dough into the oven. Dial up some Hamza El Din oud music, and wait a scant 30 minutes or so until it's brown. It will cool off quickly, so have some halloumi cheese at the ready.

Pizza

Formula

MAKES 3 MEDIUM-SIZED (12-INCH) PIZZA DOUGHS

Water	75%	630 g
Sourdough starter	5%	42 g
Bolted pizza flour	100%	840 g
Salt	2%	17 g

As any baker will tell you, pizza is first and foremost a flatbread. Without a solid, well-crafted base, all the fancy toppings in the world won't land well: it's got to begin at the beginning, with the wheat.

Most pizza in the world is pretty junky: poorly made and underbaked dough smothered in sugar-sweetened sauce and oily, cheap cheese. If you're going to spend time and money on decent ingredients, start with a high-quality pizza flour and make a naturally leavened dough.

Once again, at Hungry Ghost, we are able to count on our friends at Ground Up Grains who mill a terrific bolted pizza flour. It is in between white and whole wheat, with the stretch of the former and the depth of the latter. We leaven it with a small amount of starter (less than half of what we use for bread) and proof it overnight. Too much fermentation (whether from leavening or time) and the extensibility will be compromised; you won't be able to spin it into an even, thin crust.

These days, we mix the pizza dough twice a month and sell it frozen. For a number of years we had our own pizza service, four nights a week, and the pies were phenomenal. The bakery's limited space and production schedules just couldn't handle it, though. The

P-word is still one that makes Cheryl flinch! Here is our formula for pizza dough, plus a recipe for my favorite over the years, the Tartuffo.

Instructions

Combine the water, starter, and flour in a mixer fitted with the paddle attachment. Mix on first speed for 4 minutes. Avoid the temptation to add too much water. This dough will be considerably tighter than any of our bread doughs. Cover and let rest for 20 minutes.

Replace the paddle attachment with the dough hook; oil the hook if necessary (see How to Mix with a Stand Mixer, page 48). Uncover the dough, add the salt, and mix on second speed for 4 minutes. Again, this dough wants to be much drier than most. Put it in a very lightly oiled bin (too much oil will inhibit shaping and rolling out), cover, and give it 30 minutes in a warm spot before putting it in the fridge overnight.

Preheat the oven to its maximum heat, probably around 550°F (288°C), with your steam setup and a baking stone or parchment-lined baking sheet inside. Divide the dough into three equal pieces. Roll into tight rounds and let rest at room temperature.

Meanwhile, prep your sauce: red, white, or green. Red for tomatoes, white for a béchamel or Alfredo, and green for any kind of pesto (basil, cilantro, parsley) you might want. Then prep your veggies and your cheese.

Once the oven is up to temp, roll out the dough with a rolling pin, as evenly and roundly as possible. If you're feeling confident (or operatic!), you can gently toss or spin the dough up into the air, which, if skillfully done, will stretch the dough evenly. Some of us just get all our knuckles up under and gently exercise it, without quite going airborne. Approximate the shape of a round pizza shell, and place it on your peel. Pinch the outer rim to contain the sauce.

Spread that sauce, thinly, in a spiral from the center outward, using a soup ladle. Sprinkle your toppings and then put a final layer of cheese over it all. I find that thoughtful, well-curated choices have a better impact than grand exaggeration; less is often more.

Maneuver your construction onto the stone and don't go anywhere. Open up a beer and keep watch: 5 minutes might do it, or 10 minutes max. Allow the crust to become stiff. Contrary to popular thought, longer in a slightly cooler oven is better than singed quickly in a blast furnace. Carbonization is for self-flagellating macho dudes.

Tartuffo Pizza

Far and away my favorite, I often have no trouble eating a whole 14-inch one of these by myself. Any Rossini aria sung by Cecilia Bartoli helps my digestion.

Prepare a béchamel white sauce, beginning with a butter and flour roux, then adding milk, salt, pepper, and some grated Parmesan.

Clean and slice 3 portobello mushrooms and 1 cup shiitake mushrooms.

Shred a large handful of whole milk mozzarella (¾ to 1 cup).

Chop up some fresh chives.

Have a bottle of truffle-infused olive oil opened and at the ready.

Preheat the oven to 550°F (288°C). Stretch out the pizza dough, as described above. Spread out the white sauce, thinly, in a spiral from the center outward across the crust. Layer the mushroom slices evenly over the sauce, followed by the mozzarella. Sprinkle the fresh chives on top, and finally drizzle a smidgen of truffle-flavored olive oil over it all. Carefully transfer to the oven and bake for 5 to 10 minutes, until crispy brown. Eat it hot, but don't burn your tongue!

Folds

for the Duke
Ellington diet
take the Eight-Grain

Wouldn't you rather a silk purse than a sow's ear, even if you did eat pork? What I mean to say is, these "folds" are elegant salvage projects, taking a collapsed dough and turning it into something yummy. In the early days of the bakery, before we had our proofing trajectories dialed in, it would happen with some frequency: a loaf would overferment in the walk-in overnight and spill over the sides of the basket. As I put it on the peel, it would deflate, exactly mirroring my spirits.

Instead of throwing the dough directly into the compost (or at the wall!), I did finally learn to adapt and grab some ready-made ingredients to save the situation. Dried fruit, nuts, chocolate chunks . . . *something* good. Squash the dough flat, dock it (puncture holes with a docker, to prevent undue puffing-up), stuff half with goodies, fold it over like a Cornish pasty, and seal. Maybe spray the top and sprinkle something there. Chase the waste with taste and jam it into the jaws of defeat.

The biochemistry of overfermented dough is a field that I'd like to pretend I haven't studied, or actually camped out in, on many a working day. There is certainly a tipping point where the wished-for prized loaf will no longer be making its appearance, where the starches are sufficiently predigested by our invisible Ghost Fart friends and no Instagram close-ups will be taken. Yet, some ugly bread is in fact remarkably delicious, compensating with flavorful personality what it may lack in good looks. As any trial-and-error (and more error!) sourdough baker will tell you, there are gradations of degradation in any overproofed loaf, and you will learn to recognize and judge them. A few can be baked as is, given very gentle handling and no scoring. Some will indeed be beyond the pale, and will in fact not even bake properly, acquiring the patina of moon rocks inside the oven.

In between the acceptable-but-disappointing loaves and the trashed failures are the candidates for this creative experiment. The possibilities are endless or, rather, limited only by what you've got in the pantry or how close your corner store is. Working in a fully stocked bakery has its advantages, of course, plus we used to have an old-school grocery store directly across the street. I'd occasionally run down our steep driveway, brave a State Street crossing, and dive into Serio's Market to grab some fresh pears or whatever I could find. How we miss them!

I rarely have to make these folds anymore, having upped my dough management skills. Some customers still ask about the Chocolate Eight-Grain Fold and the French Fold. They freeze well and reheat for breakfast or dessert. Keep this trick up your sleeve . . . to pull a chocolate rabbit out of a crushed top hat.

Chocolate Eight-Grain Fold

When the Eight-Grain Bread dough has badly overproofed but not yet died a natural death, take it out of its basket and lay it on the counter. Brush off any excess flour, pat it gently, and spread it out with your hands. Use a docker to puncture the loaf and help it avoid ballooning in the oven while it bakes. Grab two large handfuls of chocolate chunks (bitter to semisweet) and cover half of the dough. Do not stint with the chocolate (or any of the other fillings), because if you don't overstuff it, the results will be underwhelming. Fold over the empty half of the dough and gently (but firmly) seal the dough together, like a bready pie-crust. You don't want that good stuff melting and oozing out.

Gently transfer to a preheated baking stone, parchment-lined baking sheet, or cast-iron skillet and bake for 40 minutes at 450°F (232°C). Pair with a nice red wine and practice saying *pain au chocolat* in French.

French Fold

When your French Bâtards have deflated and stuck to their couches, gently scrape them off (invoking Monica, Patron Saint of Patience!) and transfer to the counter. Brush off excess flour, pat down, and roll with a docker. Over one half of the dough, sprinkle a moderate amount of rubbed sage. Cover this with copious amounts of dried apricot, leaving a clear margin along the edges. Fold over the empty half and spray the top and along the edges with a water spritzer. Garnish with some sliced almonds and firmly seal all around the edges.

Bake on a preheated baking stone, parchment-lined baking sheet, or cast-iron skillet for 35 minutes at 425°F (218°C), until golden brown. Once it has cooled, serve with a chilled Champagne: you have defeated defeat!

Pear Fold

This is what to do when your Semolina-Fennel Bread has crapped out, but not quite left this mortal coil. Take it out of its basket, brush off any excess flour, and gently spread it with your hands on a clean countertop. Roll a docker over it to puncture the dough and keep it from ballooning up inside the oven. Slice up 2 to 3 pears and cover half of the dough with them. Think of it as stuffing a pie. Stretch the empty half of the dough gently over the fruit and pinch the edges sealed, spraying with water and/or rolling them up a bit if there's room. Spray the top and sprinkle with some poppy seeds.

Bake on a preheated baking stone, parchment-lined baking sheet, or cast-iron skillet at 425°F (218°C) for 35 minutes, until golden brown. Allow to cool, and serve with some cold retsina.

Savory Fold

Before we expanded and built a separate pastry department, this was one of our daily lunch offerings. It's a form of calzone, really. I would cut each standard loaf dough in half for a hand-sized fold: plenty for a filling meal.

Take a full, proofed dough of either Eight-Grain Bread or French Bâtard and brush off any excess flour. Lay it on a clean work surface and spread it out gently with your hands. Roll a docker over it. Cut it in half, horizontally, with a dough knife. Sprinkle a moderate amount of dried basil and dried onion on one half of each dough. Cover that with a slice of cheddar cheese. Pile on some sliced mushrooms and sundried tomatoes, with a second slice of cheddar to top it off. Fold the empty half of the dough over the top and seal it, using a water sprayer to help the dough edges stick together. Handle gently so that it doesn't tear (and ooze that molten cheddar out).

Place gently on a preheated baking stone, parchment-lined baking sheet, or cast-iron skillet and bake at 425°F (218°C) for 30 minutes. Eat it while warm, but not too hot! *Bon appétit!*

Going Crackers!

new day
-tare the scale
back to zero again

C rackers are easy, fast, and delicious. There are endless varieties to make, with little to none of the usual performance anxiety that accompanies proofing bread loaves. If you've never made your own before, you'll be shocked at how tasty they are, and how bland (and expensive) most store-bought versions are in comparison. Those are a triumph of packaging; yours will be the real victory.

Flour and water is all it really takes, plus some elbow grease (vegan, please!) to roll them out. The matzoh recipe here is that simple. The other recipes also have some leavening and salt, plus seeds or herbs to add accents and texture. While you can easily take our boilerplate recipe and vary the flours (glutenous or not) and other inputs, the recipes that follow are our time-tested favorites.

Matzoh

Not just matzoh, really, but Passover matzoh, which is in a category by itself. The special thing about Passover matzoh is that there is, of course, no leavening or salt (it's the bread of Exodus, and the Egyptian army was in pursuit!), and it needs to be in the oven within 18 minutes of mixing. Why 18 minutes? Because Maimonides, back in the twelfth century, calculated (correctly) that it took about that long before wild yeasts would begin to colonize the dough and render it *chametz*. Of course, in a tradition of vigorous debate, everything is debatable, including the luxury of fermentation or debate itself. Whatever the case, 18 minutes is the accepted standard, and it gives an element of playful challenge to making these crackers. While this is all in the *spirit* of being kosher, it will get you nowhere near to the real *law*. True *shmurah matzoh*, kosher for Passover, would necessitate a wheat harvest and milling that has been "watched over." Plus, a certified kitchen and oven.

Making this matzoh every spring is a fun and rewarding ritual. It could be part of any family's holiday preparations, a precursor to hiding the afikomen. I must make an admission at this point: I never measure the flour *or* the water! I mean, they were in a hurry, right? Did they bring scales and measuring cups? Far more important to me is queuing up the right music: Charles Lloyd's "Go Down Moses" (from 2010), Leonard Cohen's "Story of Isaac," and Bob Marley's "Exodus" (original studio version). They magically add up to just under 18 minutes!

Next in importance is the flour: it's got to be white spelt flour. While some believe this is one of the Biblically sanctioned grains, see the note just above on such debates. But it just plain tastes good, and its gliadin-heavy protein profile lends it to stretching out easily into thin crackers. Don't use whole or bolted spelt: white will get you to the seder table on time. Grab a large bowl, fill it halfway with white spelt flour. Have a large pitcher of clean, cold water at hand, plus a plastic dough scraper, silicone spatula, dough knife, extra flour for kneading, rolling pin, wooden peel or sheet of parchment paper, long-handled offset spatula, and pastry docker (Miriam must have brought those along!).

Preheat the oven to 425°F (218°C), with a baking stone inside (if using).

Cue the music, start your timer, and begin pouring water into the bowl and mixing with the spatula. Don't flood the flour, but incorporate it as quickly and dryly as you can, graduating to the dough scraper when needed. Add water if there's loose flour, and if not, then dump the contents of the bowl onto a floured countertop and start kneading it, smoothing it out and working as much flour as you can into it. Your hands will get sticky and gummy, but don't wash them! Rub them together and keep kneading. Knead until the dough gets less and less sticky. You'll be into the second song by now.

Use the heels of your hand to get maximum leverage and pressure, folding the dough over onto itself, repeatedly. Still dipping your hands into the spelt flour and sprinkling it around, get a good, stiff ball of dough. Cut it into quarters. If you have a wooden peel that will transfer the cracker dough to a preheated baking stone, roll it out on that. Otherwise, use a sheet of parchment paper, which will then go onto a baking sheet.

Roll the dough out, first in one direction and then in another. Keep the flour flying, to keep things from sticking to each other. Use the offset spatula to unstick it from below. When it is as thin as you can get it, and loose, brush it off quickly and score it with the docker (that will keep it from bubbling up while baking). Put it in the oven and start rolling out the next cracker.

When the timer goes off, don't stop—just know that that was Maimonides, looking over your shoulder. Keep going! Watch the crackers in the oven; they will brown up within 8 minutes or so, and you don't want them burning. Once they're decidedly stiff, not floppy, and browning up, put them on a rack to cool. You've just had your 40 seconds in the Wilderness. Most matzoh, taste-wise, is indistinguishable from the boxes they come in. This will have the flavor of liberation.

Sesame-Spelt Crackers

Ingredients
MAKES 2 LARGE (40 OZ) BATCHES

White spelt flour	1 kg
Salt	40 g
Sesame seeds	142 g
Sourdough starter	240 g
Water	1 kg

These are the nonholiday, year-round, salted and leavened version of matzoh. Sesame seeds add a further textural dimension. As crisp and satisfying as any potato or corn chip, they have absolutely zero of the problematic oils associated with those snacks. Give them a try.

Instructions

Mix the flour, salt, and sesame seeds together in a large bowl, big enough to hold at least twice the volume of these ingredients. Blend them together with your fingers in a swirling motion. Make a little crater in the middle and pour in the starter. Begin to drizzle in some water while stirring with a silicone spatula. Continue stirring and pouring until the water is used up and incorporated. Use a plastic dough scraper to unstick the dough from the sides of the bowl, and keep mixing. If it needs more water (depending on the protein levels of your particular batch of flour), then add a bit at a time. Otherwise, when the dough becomes hard to maneuver in the bowl, dump it on a clean working surface (wood is often best), ditch the spatula, and get your ungrubbied little hands in there.

Some refer to this as the "therapeutic" action of kneading: they take their frustrations out on the dough, conjuring up some nemesis to motivate their force. Personally, I think it's bad for food to be prepared with deflected passion of any kind, rather than conscious intent. Resentment depletes, while generosity nourishes.

This is needful kneading, then: work the dough, folding it in on itself and incorporating more flour as needed. The goal is to get a consistent, tight, somewhat springy and only vaguely tacky ball of dough. If you find your hands and forearms straining, cover the dough with a food-grade plastic bag or a clean tea towel and take a break. Both the gluten strands and your muscles need to relax, and both will benefit from a 20-minute rest.

Finish off the kneading when you can. You can scarcely hand-mix it too tightly. The drier it is now, the less it will stick when you're rolling it out later. Flour the mixing bowl and put the dough back in it, covering with the plastic bag or towel. Put it in a warm spot for a couple of hours. Due to the inclusion of the starter, this dough will not store indefinitely, nor even do well overnight. If it proofs for too long, it will not hold up to vigorous stretching or rolling out, and will tear.

Preheat the oven to 425°F (218°C), with a baking stone (if available) inside.

Divide the dough into four pieces. On a floured surface, roll out the dough as thinly and evenly as possible, using your largest rolling pin. Use an offset spatula to dislodge it from your working surface and a dry pastry brush to clean off extra flour. Roll the docker over it when you're ready to load it into the oven. Transfer it onto the pizza stone or a sheet pan lined with parchment paper. I then use a small fluted (or crimped) pastry cutter to perforate it lengthwise: this makes for cleaner cracker edges once they're baked.

Bake for 8 to 10 minutes, keeping a close eye on it. Don't get lost in some podcast; focus! Keep your eyes on the prize inside the cracker box, which turns out to be . . . some *very* good crackers! If one part browns up before the rest, then get out your pizza cutter, slice off that part, and let it cool. As long as the crackers are stiff and not floppy, they are good to go. They will finish hardening while they cool.

They cool quickly. Is your Chablis cold yet? Cheese at room temp? Miles Davis record ready for the needle?

Khorasan Kalonji Crackers

Ingredients
MAKES 2 LARGE (40 OZ) BATCHES

White khorasan flour	1 kg
Salt	40 g
Kalonji seeds	70 g
Sourdough starter	240 g
Water	1 kg

Like spelt, khorasan is an older cousin of wheat, with a somewhat different protein makeup. Named for its land of origin in present-day northeastern Iran, it is a hard durum wheat with a beautiful yellow tinge. Many Americans are familiar with a trademarked version of it called Kamut, though that sports a supposed Egyptian provenance. Also like spelt, khorasan has good extensibility, so I recommend using white rather than whole wheat. Unlike spelt, khorasan also has quite a bit of elasticity and can handle a remarkable amount of hydration, both in cracker and bread dough.

Kalonji seeds are little flavor bombs that are used throughout Central Asia and beyond. They are also known as nigella, charnushka, black cumin, or Roman coriander. They are the seeds of a flowering plant known variously as love-in-a-mist or devil-in-the-bush—depending on how you feel about the taste? Or perhaps it's not an either/or, but rather love-in-a-mist-with-a-devil-in-the-bush.

Spicy stuff, anyhow. We use half as much as we do sesame seeds in the Sesame-Spelt Crackers. Otherwise, this cracker has similar proportions.

Instructions

Mix the flour, salt, and seeds together in a large bowl, big enough to hold at least twice the volume of these ingredients. Swirl them around with your fingers. Make a little crater in the middle and add the starter. Slowly pour in the water as you mix with a silicone spatula, eventually graduating to a stiffer and wider plastic dough scraper. Add whatever water seems necessary, then turn everything out onto a clean working surface and knead until it is cohesive, uniform, and barely sticky at all. Use extra flour as needed. Flour the bowl, return the dough to it, and cover. Let it rest for a couple of hours in a warm place.

Preheat the oven to 425°F (218°C), with a baking stone (if available) inside.

Divide the dough into four manageable pieces and roll them out with a rolling pin as thinly and evenly as possible. A long, offset metal spatula will help you get the dough unstuck from the table. When it is all stretched out, roll a docker all over it and, if you have one, a pastry roller.

Slide it onto the pizza stone with a peel or transfer to a parchment-lined sheet pan and bake for a well-watched 8 to 10 minutes. Remove when it is barely browning but stiff (rather than floppy). Allow to cool before gorging.

Rye & Onion Crackers

Ingredients
MAKES 2 LARGE (40 OZ) BATCHES

Bolted rye flour	500 g
White bread flour	500 g
Salt	40 g
Dried onion (crumble it!)	56 g
Sourdough starter	240 g
Water	1 kg

Rye flour, like just about everything else, grows in complexity the closer you get to it. There's more than one generic type, to start with. There are aromatic, flowery varieties. There is winter- and spring-planted rye (just like wheat), and there are varieties from all over Central Asia and Europe. Ground Up Grains mills a Danko rye grown in western New York State, and it is delicious. Though there isn't much wheat grown around the Pioneer Valley where we live, there *is* plenty of rye, generally planted as a cover crop to fix nitrogen into the soil. I love spotting its blue-green blur as I bike past fields in the late spring. I would pair this cracker with a soft goat cheese and a pint of local ale (not too hoppy). The dried onion in this recipe can burn if the oven's too hot or if the crackers are in there a few moments too long, so please be careful!

Instructions
Mix the flours, salt, and dried onion together in a large bowl, big enough to hold at least twice the volume of these ingredients. Swirl

them around with your fingers. Make a little crater in the middle and add the starter. Slowly pour in the water as you mix with a silicone spatula, eventually graduating to a stiffer and wider plastic dough scraper. Add whatever water seems necessary, then turn everything out onto a clean working surface and knead until it is cohesive and uniform and barely sticky at all. Use extra flour as needed. Flour the bowl, return the dough to it, and cover. Let it rest for a couple of hours in a warm place.

Preheat the oven to 400°F (204.5°C), with a baking stone (if available) inside.

Divide the dough into four manageable pieces and roll them out with a rolling pin as thinly and evenly as possible. A long, offset metal spatula will help get it unstuck from the table. When it is all stretched out, roll a docker all over it, followed by a pastry roller if you have one.

Slide it onto the pizza stone with a peel or transfer to a parchment-lined sheet pan and bake for a well-watched 8 to 10 minutes. Remove when it is barely browning but stiff (rather than floppy). Cool before biting!

Semolina Sage Crackers

Ingredients
MAKES 2 LARGE (40 OZ) BATCHES

White durum semolina	500 g
White bread flour	500 g
Rubbed sage	115 g
Salt	40 g
Sourdough starter	240 g
Water	1 kg

Semolina is a particularly coarse milling stream from particularly hard durum wheat, so it makes for a high-protein, large-grained golden flour. It is primarily used in couscous and pasta, though we blend it with white flour for loaf breads, flatbreads, and crackers. Keep some in your baking pantry! Also, rubbed sage, my favorite. I often gather fresh sage leaves in the garden to chop up (or occasionally use whole), and I'm always taken aback at how much stronger dried herbs are. That's why witches always have them hanging from their kitchen rafters! Their power concentrates as they wither and crone (both the herbs and the witches, I guess, but it wouldn't be sage to conjecture any further).

Instructions

Mix the flours, sage, and salt together in a large bowl, big enough to hold at least twice the volume of these ingredients. Swirl them around with your fingers. Make a little crater in the middle and add the starter. Slowly pour in the water as you mix with a silicone spatula, eventually graduating to a stiffer and wider plastic dough scraper. Add whatever water seems necessary, then turn everything out onto a clean working surface and knead until it is cohesive, uniform, and barely sticky at all. Use extra flour as needed. Flour the bowl, return the dough to it, and cover. Let it rest for a couple of hours in a warm place.

Preheat the oven to 425°F (218°C), with a baking stone (if available) inside.

Divide the dough into four manageable pieces and roll them out with a rolling pin as thinly and evenly as possible. A long, offset metal spatula will help get it unstuck from the table. When it is all stretched out, roll a docker all over it and, if you have one, a pastry roller.

Slide it onto the pizza stone with a peel or transfer to a parchment-lined sheet pan and bake for a well-watched 8 to 10 minutes. Remove when it is barely browning, but stiff (rather than floppy). Allow to cool. The smell of the sage always make me reach for apricot jam, though sometimes it's right next to the cream cheese in the fridge, and then . . .

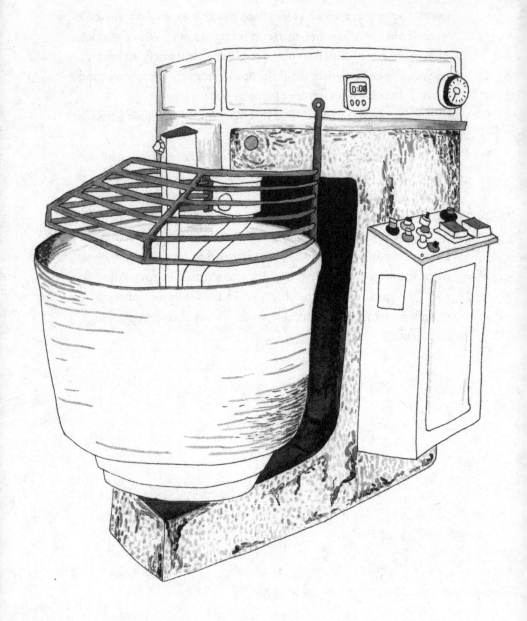

Scones

Rosemary leaves
right through the
dough-er

A true scone is a dry and crumbly thing, and not too sweet. Its contents may vary, as will its pronunciation (skON, or skOWn) but keep in mind: There's a Scottish town, capital of an ancient kingdom, that goes by the same name, and in the year 1296, Edward I of England stole the huge Stone of Scone to use for his coronation. Every English monarch since has had it neath their throne, if only (lately) on loan. That's a lot of history for a little breakfast treat!

On the wall of our bakery's pastry department are these notes for scones and biscuits:

Keep ALL the ingredients as cold as possible. (Don't freeze butter though; it doesn't break down as nicely).

Do not overmix, and this includes the butter. There should be visible chunks of butter in the final dough. How long the mixing takes will vary according to butter cube size and weather. Adjust accordingly.

Add liquid slowly, and stop mixing once the dough comes together. You may need to mix in a little by hand, or distribute wetter pieces with drier pieces. Do this by hand, not in the mixer.

Handle the dough as little as possible. It's okay to reroll, but do *not* wad the dough; stack pieces on top of one another.

Work quickly so the dough isn't sitting out too long. We want the chemical reaction to happen in the oven, not while the dough is sitting on the counter. Brush off excess flour.

Bake until golden. Do not underbake; pale pastries are sad.

Rosemary Walnut Scones

Ingredients
MAKES 8 LARGE (5½ OZ) SCONES

All-purpose flour, chilled	310 g
Pastry flour, chilled	200 g
Sugar	¼ cup (50 g)
Baking powder	1 tablespoon (12 g)
Baking soda	1 teaspoon (6 g)
Salt	1 teaspoon (6 g)
Butter, chilled, then cubed or sliced	227 g (2 sticks)
Dried rosemary	4 teaspoons (5 g)
Walnuts	1 cup (117 g)
Buttermilk	1 cup (242 g)
Heavy cream	1 tablespoon (14 g)

Maple-buttermilk glaze: 1 tablespoon buttermilk (14 g) whisked with ¼ cup (78 g) maple syrup

---

Bikepacking in Scotland consists of constantly gasping at the landscape and constantly searching for carbohydrates. It seems the Romans brought both wild rosemary and walnuts to Britannia, and they must have deposited them at Hadrian's Wall. I am nuts about the scones at Hadrian's Wall.

Instructions

Preheat the oven to 375°F (190.5°C).

Combine the flours, sugar, baking powder, baking soda, and salt in a large bowl and set aside in the freezer for at least 10 minutes

Transfer the dry ingredients to a mixer fitted with the paddle attachment, add the butter, and mix on low speed until the butter begins to break down into smaller pieces. Now, add the rosemary and walnuts.

Combine the buttermilk and cream. Slowly add the buttermilk mixture at moderate speed until it begins to come together. Once combined, turn out onto a floured surface and make 2 to 4 folds by cutting the dough in half, then stacking one half on top of the other and pressing down. Roll out the stack to an 8-inch square. Trim the edges straight with a dough knife. Cut down the center and crosswise to create four 4-inch squares, then cut diagonally across each square to create eight triangles. Place the scones on a parchment-lined baking sheet.

Brush each scone with the maple-buttermilk glaze (whisked together for a loose, but not watery consistency) and bake for about 14 minutes, or until they look golden-brown delicious: crispy on the edges with a center that has give but is not mushy. Allow to cool for 2 minutes, then brush with the glaze again.

Fig & Honey Scones

Ingredients
MAKES 8 LARGE (5½ OZ) SCONES

Bolted or white spelt flour, chilled	510 g
Baking powder	1 tablespoon (12 g)
Baking soda	1 teaspoon (6 g)
Salt	1 teaspoon (6 g)
Butter, chilled, then cubed or sliced	227 g (2 sticks)
Buttermilk	1 cup (242 g)
Honey	3 tablespoons (63 g)
Heavy cream	1 tablespoon (14 g)
Dried figs, roughly chopped	1 cup (170 g)

Egg wash: 1 egg yolk whisked with a splash of heavy cream

My great-grandfather Stevens was born in Penzance, Cornwall, though he was not a pirate (so far as I know). Cornwall is the Celtic pointy bit in the lower left-hand corner of England, jutting out into the Channel. The Coastal Path there winds along the cliff tops, and every once in a while you'll stumble onto a shop serving Cornish cream tea. Not just tea, of course, but also clotted cream and strawberries smothering a scone. Like this one.

This is the same basic recipe as the Rosemary-Walnut Scones, but it subs out honey for sugar, spelt flour for the all-purpose and pastry flours, dried figs for the rosemary and walnuts, and an egg wash for the glaze. It will crown your day.

Instructions

Preheat the oven to 375°F (190.5°C).

Combine the flour, baking powder, baking soda, and salt in a large bowl and set aside in the freezer for at least 10 minutes (or up to 2 years). Transfer the dry ingredients to a mixer fitted with the paddle attachment, add the butter, and mix on low speed until the butter begins to break down into smaller pieces.

Combine the buttermilk, honey, and cream. Add the dried figs. Slowly add the buttermilk mixture at moderate speed until it begins to come together. Once combined, turn out onto a floured surface and make 2 to 4 folds by cutting the dough in half, then stacking one half on top of the other and pressing down. Roll out the stack to an 8-inch square. Trim the edges straight with a dough knife. Cut down the center and crosswise to create four 4-inch squares. Cut diagonally across each square to create eight triangles. Place the scones on a parchment-lined baking sheet.

Brush each scone with the egg wash and bake for 14 minutes, or until they look golden-brown delicious, with crispy edges and soft but not mushy centers. Allow to cool, and enjoy!

Blueberry Cornmeal Scones

Ingredients
MAKES 8 LARGE (5½ OZ) SCONES

Pastry flour, chilled	200 g
Cornmeal (mix of coarse and fine)	170 g
All-purpose flour, chilled	140 g
Sugar	¼ cup (50 g)
Baking powder	1 tablespoon (12 g)
Baking soda	1 teaspoon (6 g)
Salt	1 teaspoon (6 g)
Butter, chilled, then cubed or sliced	227 g (2 sticks)
Buttermilk	1 cup (242 g)
Heavy cream	1 tablespoon (14 g)
Blueberries, fresh or frozen	1 cup (190 g)

Egg wash: 1 egg yolk whisked with a splash of heavy cream

Blueberries from Blue Hill, hand-raked with a view of the bay and Mount Desert Island in the distance: that's my "Maine" inspiration for these scones. With some Maine Grains cornmeal milled in the old jailhouse in Skowhegan, grown at Liberation Farms! That's the way life should be.

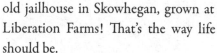

A mixture of coarse polenta grits and fine-ground cornmeal is ideal for this recipe. Fresh blueberries are too, but frozen will do the trick if that's all you have.

Instructions

Preheat the oven to 375°F (190.5°C).

Combine the pastry flour, cornmeal, all-purpose flour, sugar, baking powder, baking soda, and salt in a large bowl and set aside in freezer for at least 10 minutes. Transfer the dry ingredients to a mixer fitted with the paddle attachment, add the butter, and mix on low speed until the butter begins to break down into smaller pieces.

Combine the buttermilk and cream. Add the blueberries. Slowly add the buttermilk mixture at moderate speed until it begins to come together. Once combined, turn out onto a floured surface and make 2 to 4 folds by cutting the dough in half, then stacking one half on top of the other and pressing down. Roll out the stack to an 8-inch square. Trim the edges straight with a dough knife. Cut down the center and crosswise to create four 4-inch squares. Cut diagonally across each square to create eight triangles. Place the scones on a parchment-lined baking sheet.

Brush each scone with the egg wash and bake for 14 minutes, or until golden-brown delicious, with crispy edges and a soft but not mushy center.

Not Alone

as Jesus said
to Judas, better
bread than dead

The word made flesh, the recipe made real, John Barleycorn flayed into whiskey: it's all of a piece. We sacrifice some to the gods, some to the middleman, and we keep a little for ourselves. Never bread *alone*: if not anointed with olive oil, then enriched with butter and eggs. Boiled as bagels, rounded into biscuits, rolled flat into pasta. Muffins, shortbread, fruit bars, and barley cakes: this is our flour'ed litany. No need to pray for them to taste good and nourish. Your good intentions will suffice.

Bagels

Formula
MAKES ABOUT 15 (100 G) BAGELS

Egg	1 large egg per kilo of flour	
Honey	4.5%	38 g
Olive oil	3%	25 g
Water	64%	538 g
Starter	12%	100 g
Strong white bread flour (14% protein)	100%	840 g
Salt	3%	25 g

Additional honey, as desired, for the boiling water
Sesame and/or poppy seeds, for coating

Montrealers have a kind of underdog chauvinism when it comes to bagels (and hockey): "Those poor fools elsewhere" (even New York and Boston), "they don't even know what real bagels are!" They may look vaguely like bagels, but what you're used to are in fact bread doughnuts. A real bagel is skinny, chewy, and a little sweet. Boiled in honey water, then baked on a wood-fired hearth. Eaten while still warm, in the steamy doorway of the shop; maybe it's the middle of the night after a show, on a side street in the Plateau.

As with New Haven pizza mavens, there are strong feelings about the best place to go, though there are in fact only two choices, five blocks apart: Saint-Viateur and Fairmount. Both are tiny, bake only bagels, and are open 24 hours a day. You could walk briskly, bagel in hand, from one to the other, just to compare, and the first bagel would still be warm—though I wouldn't recommend it. Someone may take offense.

Stay a while at either to see the process, since it's quite a show: slicing off hunks of white dough on the wooden table, quickly rolling out thin snakes, then looping them 'round the back of the hand so each end seals up under a rolling palm. Mere seconds pass as the pile of dough rings heaps up. A short while later, they're dipped into a huge boiling kettle of water sweetened with honey. After bobbing around in the bubbling cauldron for a few minutes, they're fished out and coated in sesame or poppy seeds.

Then the long, narrow sheeba (bagel peel) comes into play, like some baker's version of Charon's boat paddle, ferrying the bagels to their fate beside the fire. The oven is a wide rectangular opening in the back wall, no door on it at all, with a live, well-tended fire of flames and embers off to one side. The oven's so deep that the long sheeba barely touches the back as it deposits its dozens down. Long rows of gradually baking beauties are maneuvered further and further from the licking heat, and halfway through, they're flipped completely over. With one last flourish, the sheeba collects its baked-off brood and with a flick of the wrist, the baker sends them flying in an acrobatic arc into a basket on the floor.

This is indeed where I first fell in love with baking, with the every-day-is-Christmas feeling of feeding people in a warm space with the snow piling up outside. There is also a village-within-the-city cosmopolitan vibe that this edible bracelet connotes—an exile's gift to their newfound neighbors. It's a Polish king's stirrup transfigured into food, a baby's teething toy garnished with a schmear of cream cheese. It is a lot of things to a lot of people, as only comfort food can be.

Without my own sheeba or rectangular black oven or immense bagel boiler, I will never be able to quite fulfill my first baking dream. But our version comes pretty close (depending on who in my family you believe). Our bagel dough is essentially our Challah dough, used the same day as mixing (without an overnight bulk proof). Shaped and boiled and baked, this naturally leavened version lasts a little longer than the commercial kind, though they will be a little less sweet (and saltier). Don't forget to listen to Leonard Cohen's cover of "Un Canadien Errant," replete with mariachi band.

Instructions

Beat the egg in a bowl, then whisk in the honey and olive oil. Combine the water, starter, and flour in a mixer fitted with the paddle attachment, then add the egg mixture. Mix on first speed for 4 minutes, enough to incorporate well. Cover and let rest for 20 minutes.

Replace the paddle attachment with the dough hook; oil the hook if necessary (see How to Mix with a Stand Mixer, page 48). Uncover the dough and add the salt right away, while on first speed, then gear up to second speed for 4 minutes. Keep this dough as tight as possible, without adding additional water. Transfer to an oiled bin, cover, and let it sit for about an hour.

Preheat the oven to 425°F (218°C), with a baking stone (if available) inside.

Turn the dough out onto a lightly floured work surface. Cut off a hunk about half the size of your fist and roll it out into a skinny snake the length of your outstretched hand, pinky to thumbnail. Now whip it around the back of your hand, holding on to one end between your thumb and index knuckles. The two ends of the snake will overlap in your palm. Then press the mouth and tail together on the table and roll your hand back and forth, sealing them together in a doughy ouroboros. Don't let this ring unravel; seal it well and swivel it around the middle of your hand like Mardi Gras beads, evening out the circle. Roll out all the dough, not minding how unique each bagel might appear. Keep the holes on the wider, larger side, and the rings skinny. Dust with flour to keep them from closing up or sticking.

Get your largest pot of water on to boil—a full, roiling boil—with a generous amount of additional honey poured in. Gently deposit two or three bagels into the pot—not from way up above the water, or it will splash at you. Touch the bottom end of each bagel to the surface of the water and then let go (something I learned dropping drumsticks into oil at KFC, age 15!). Prod them in their tumultuous drowning till they float back to the top, and then skim them out. Put them aside without drying them off, and do the rest. Take the boiled and still-wet rings and dip them in a dish of sesame or poppy seeds to thoroughly cover. The honey will help glaze the bagels and the seeds will insulate the crust in the hot oven.

Slide your bagels onto the baking stone or a parchment-lined baking sheet and bake for 20 to 25 minutes, until nicely browned. Flip them over for a final 5 minutes or so. Let cool on a rack for just a few minutes, then break out the cream cheese and the Fin du Monde triple blonde beer.

Buttermilk Biscuits

Ingredients

MAKES 8 LARGE (6½ OZ) BISCUITS

All-purpose flour	255 g
Pastry flour (bolted)	255 g
Baking powder	1 tablespoon (12 g)
Sugar	2 teaspoons (8 g)
Salt	1 teaspoon (6 g)
Baking soda	1 teaspoon (6 g)
Chilled butter, cubed or sliced	1–1½ cups (242 g–363 g)
Buttermilk	3 cups (726 g)

There's upwards of twenty natural cascades throughout New York, New Jersey, and Pennsylvania named Buttermilk Falls. If you dropped your biscuits into one of them, where would your breakfast be?

Good buttermilk is the secret to these biscuits, as is bolted wheat pastry flour, if you can find it. In the summertime, these are very popular as a pedestal for fresh strawberries and cream!

Instructions

Combine the flours, baking powder, sugar, salt, and baking soda in a large bowl. Add the cubed butter and place in the freezer for 10 minutes to 1 hour. Transfer to a mixer fitted with the paddle attachment and mix until the butter is mostly broken down into hazelnut-sized pieces. Start adding the buttermilk and keep mixing until the flour is mostly moistened and combined. You can fold in any loose bits at the bottom by hand.

Preheat the oven to 375°F (190.5°C).

Flour your table and roll the dough out with a rolling pin to about 1¼ inches thick. Use a 1-cup measure to cut perfect rounds out of the flattened dough. Reroll as needed by piling the left-over trimmings and then pressing down—don't wad the dough. Try to handle it as little as possible! Brush off any excess flour and place on a parchment-lined baking sheet.

Bake for 14 minutes. You want the biscuits to be pretty deep golden.

Egg-in-a-Biscuit

My daily (second!) breakfast at work is this enhanced version of the noble biscuit—one that adds vital fats and protein. Mix and prep your biscuits as above; then, once the raw dough rounds are on the parchment paper, cut out an inner round (1 inch in diameter) at the center of each biscuit. Cut all the way through and remove the center. Then crack a raw egg into each hole. Sprinkle a pinch each of salt and pepper on top, followed by a smattering of shredded cheddar cheese.

Bake as above, at 375°F (190.5°C), for 14 minutes. The egg might ooze out a bit, but don't worry or overcook it; it's better with the yolk on the looser side.

Bran Muffins

Ingredients
MAKES 12 MUFFINS

Wheat bran	90 g
Yogurt	1 cup (227 g)
Sunflower oil	½ cup (114 g)
Large eggs	2
Barley malt syrup	½ cup (186 g)
Baking soda	1 teaspoon (6 g)
Baking powder	1 teaspoon (4 g)
Salt	½ teaspoon (3 g)
Dried cranberries (unsweetened!)	1 cup (121 g)
Grated lemon zest	2 teaspoons (4 g)

Muffins are not meant to be pseudo-cupcakes. Icing, as far as I'm concerned, is a penalty in both hockey *and* baking. The first outstanding bran muffin I recall eating was at Trident Booksellers & Cafe in Boulder, Colorado, in 1985. It was that good. It was also the first time I saw coffee afficionados line up for a barista, but then everyone in Boulder seemed to have time to hang around back then.

Hanging around was not quite was I was doing when I set out to create my own version of these muffins, eleven years later. Up at the crack of dawn with a baby in a sling, making breakfast before the four-year-old woke up too. Determined to deny my kids refined sugar (and television!) for as long as possible, barley malt became my favorite substitute. We all eventually recovered from my strict principles.

Instructions

Preheat the oven to 375°F (190.5°C). Oil 6 cups of a standard muffin tin.

Mix the bran and yogurt together in a large bowl, and let that sit for 10 minutes. In a separate bowl, beat the oil, eggs, and barley malt syrup together, then stir that into the bran-yogurt mixture. Sift the baking soda, baking powder, and salt together thoroughly, then gently add that to the mixture, along with the dried cranberries and lemon zest. Do not overmix.

Spoon the batter into the oiled muffin cups, about two-thirds full, distributing the batter evenly. Bake for 20 minutes, or until the thinnest edges look crispy and enticing. Share with the nearest toddler.

Maple Shortbread

Ingredients
MAKES 16 SHORTBREAD SLICES

Butter, softened	340 g (3 sticks)
Maple sugar	200 g
All-purpose flour	450 g
Salt	1 teaspoon (6 g)

1 large egg, whisked
Sea salt, for sprinkling

Maple sugar is, of course, the sweetener that we're blessed with here in the Northeast, another gift that First Nations communities shared with European settlers. Growing up in Quebec, it was impossible to avoid the seasonal significance of what we called sugaring-off in late winter and early spring. My best memories of those times are of warm maple taffy hardening on clean picnic-table snow. A deftly twirled twig would roll the congealing magic into a forest lollipop. Hardwood groves, sap buckets hung, snowshoe tracks, and steam pouring out of the sugar shack: that's what maple syrup means to me.

Shortbread is of Scottish origin, linked in popular imagination to Mary, Queen of Scots and her alliance (culinarily, militarily, and linguistically) to the French. As it so happens, the forest preserve that I was taken to as a child is on the campus of Macdonald College, in the town of Sainte-Anne-de-Bellevue, across the river from the Mohawk village of Kanesatake. Maple Shortbread, indeed.

Instructions

Preheat the oven to 350°F (176.5°C).

On a clean work surface, paste together the butter and maple sugar by sprinkling the sugar over the spread-out, softened butter and folding the mass over on itself repeatedly. Add the flour and salt and repeat the slow incorporation. Roll into logs about 2½ inches across and 10 to 12 inches long.

Slice each log into 16 slices and place on a parchment-lined baking sheet. Brush with the whisked egg, crosshatch with a fork, and sprinkle with sea salt. Bake for 12 to 16 minutes, until deep golden brown.

Fruit Bars

Ingredients
MAKES 16 BARS

Dried figs	900 g
Apple juice (unsweetened)	3 cups (753 g)
All-purpose flour	400 g
Pastry flour (bolted, if available)	225 g
Rolled oats	400 g
Vegan butter	510 g
Sugar	400 g
Desiccated, shredded coconut	140 g
Salt	¾ teaspoon (5 g)
Baking soda	¾ teaspoon (5 g)

A man walks into a Fruit Bar and says, "Wow, this is delicious!" Someone I know puts a dollop of whole milk yogurt on this (otherwise vegan) confection and swears it tastes "just like friggin' cheesecake." Not a fig-ment of your imagination. We use sunflower-derived Earth Balance vegan buttery spread for this recipe.

Instructions
Preheat the oven to 350°F (176.5°C).

To make the filling, combine the figs and apple juice in a large saucepan. Bring to boil, then turn down the heat and simmer until reduced to a thick paste, about 20 minutes.

To make the crumb/crust, combine the flours, rolled oats, vegan butter, sugar, coconut, salt, and baking soda in a large bowl and mix until just integrated. Press half of the mixture into a parchment-lined 13 × 18-inch baking pan. Spread the filling on top. Loosely crumble the rest of the crumbs on top (like streusel). Bake for 40 minutes, until they are toasty brown.

Cyclist Repair Kit

Ingredients
MAKES A DOZEN LARGE ROUND TREATS

Flax meal	1 cup (104 g)
Apple juice	1 cup (251 g)
Whole malted barley seeds	1 cup (105 g)
Warm water	1 cup (240 g)
Tamari-roasted cashews, broken into pieces	1 cup (120 g)
Chopped dates	1 cup (160 g)
Malted barley flour	1 cup (100 g)
Sesame seeds	

Spring, summer, and fall, I commute to work by bike, 10 miles each way. On weekends I'll go for longer rides, on either gravel roads or dirt tracks. In the winter, if there's snow, nothing's better than a cross-country ski glide through the woods. Often the hardest part is packing enough protein to keep from "bonking." There's only so many pseudo-healthy candy bars my stomach can stand!

So, I developed an alternative that works well for me, with lots of my favorite ingredients, including barley. Last summer, on my annual cycling trip, I took along eight of these babies to southern England, and found myself rolling through acres and acres of wind-shook barley fields while navigating old Roman roads: definitely heaven.

You'll notice there's no white sugar in this recipe. I certainly got my share of that stuff early in life, raised in a family of sweet tooths, but it gives me a headache, without fail. Of course, sugarcane gave the *whole world* a diabetic shock, thanks to plantation slavery, and that's yet to be healed. Barley, on the other hand, is a remarkable source of complex sweetness, brought to the fore when the grain is malted. Sprouting the seed tricks its sugars into play; then they're toasted, locking them in place. Another astounding thing about barley is that its soluble fiber, known as beta-glucans, actually lower your cholesterol!

Instructions

In a small bowl, soak the flax meal in the apple juice for 30 minutes. At the same time, in another small bowl, soak the malted barley seeds in the warm water for 30 minutes.

Preheat the oven to 375°F (190.5°C).

Combine the soaked mixtures in a large bowl. Using a silicone spatula, fold in the tamari-roasted cashews, chopped dates, and malted barley flour. Add extra apple juice or malted barley flour as needed to get a moist and pliable dough.

Scoop the dough (I use an ice cream scoop) and drop into a bowl of sesame seeds, coating the ball and then flattening it out into a patty between your palms. I use an Uzbeki bread stamp to compress it a little further. Place on a parchment-lined baking sheet and bake for 25 minutes or until brown and firm. Let them cool completely, then pump up your tires, put on your helmet, and wait a few miles to start nibbling away.

Fresh Sourdough Pasta

Ingredients
MAKES 1 LARGE SERVING OF PASTA

Semolina, spelt, or khorasan flour	1 cup (130 g)
Sourdough starter or starter discard	¼ cup (57 g)
Water	½ cup (120 g)
Salt, for the pot	

If you can talk, you can sing. If you can walk, you can dance. And if you've got ingredients out to make bread dough, you can easily make great pasta. Making fresh pasta can seem like a big production or a chore sometimes, but honestly, if you've made a floury mess of your kitchen already and still have to wait till the next day to bake, then seize the fettuccine moment!

This is supremely simple pasta, with no egg, no salt, and no oil. Just water, flour, and sourdough starter.

Why the starter? Well . . . why not? It bestows flavor and digestibility to noodles, both of which are sorely lacking in the store-bought, boxed-up versions we're all accustomed to. Instead, these are delicate, lively, and fast-boiling: You'll barely have time for the checkered tablecloth and the Chianti candle holder (BTW, that bottle's called a *fiasco!*).

I recommend using spelt, khorasan, or semolina durum flour for fresh pasta. They're characteristically suited to stretching out into thin strips (thanks to their distinctive protein profiles), and besides, they taste really good. Use up extra or discard starter that you have left over from making bread. This is an informal recipe, a *strategy* really. Don't gear up too much to do it. Leave the fancy pasta machine in the cupboard. Just whip it up and roll it out, cut it into strips, and boil. All recipe amounts are approximate.

Instructions

Put the flour in a large bowl and carve out a small crater in the middle. Pour in the starter and then the water, all the while stirring with a silicone spatula. As the mix stiffens, use a plastic dough scraper to scour the sides of the bowl and fold any remaining, unincorporated flour into the dough. Drizzle in a little more water, if necessary. Unlike bread dough, you want this dough to be as dry and tight as possible (as you would for crackers). Turn out onto a floured work surface and knead it with your fingers and the heels of your hands. When you've worked in as much extra flour as you can, put it back into the (lightly floured) bowl and cover. Allow it to sit aside for at least 30 minutes (for the gluten to relax) but no longer than 1 hour (so that fermentation won't start to disintegrate those same gluten strands).

While you're waiting, get a large pot of water up to a boil and prep whatever kind of sauce you'll want on top: maybe a marinara or Alfredo or pesto. Or you could just have butter and Parmesan. Set up your colander and serving bowl, too, so you'll be *really* ready when the noodles are. Turn the water down to a simmer.

Again, turn the dough out onto a floured work surface. Grab your heaviest rolling pin and get to work, bearing down evenly and spreading the dough outward in one direction and then the other, all the while sprinkling extra flour to keep things from sticking. Fold it back on itself, rotate 90 degrees, and start again. Repeat this step 4 or 5 times, resting your arms as necessary (and allowing the gluten to do the same). Once you've thinned it out as much as possible, working it back and forth and a little side to side, you'll have a jagged rectangle of sorts. Use an offset spatula to dislodge it from the countertop, generously sprinkling additional flour on, around, and underneath it.

It's time to cut the dough into thin strips, maybe ¼ inch wide. You can use a rolling pastry cutter or just a paring knife. For consistency of cooking, try to make the ribbons an even width, with cuts as parallel as possible. Cut them lengthwise, as well, to about 7 or 8 inches, for ease of handling. If you've got a rack of some sort at hand, drape them gently; otherwise, lay them on some parchment paper. Bring the pot of water back to a roiling boil, making sure to add a generous tablespoon of salt (because the dough has none!). Once it's bubbling, gently slide the noodles in, stirring lightly to separate them. Keep the flame on high, since this part takes only 3 to 4 minutes. Stand by, still stirring, and preparing to teeth-test the results. As soon as you get that al dente bite-through, ferry the whole steaming production over to the sink and drain it through a colander. Shake any excess water through, then plop the pasta onto your favorite pasta platter, top it with sauce, and get your company to sit down *pronto* and tuck in.

Leftover Bread

Yesterday's yeast
turns into
all of tomorrow's toast

In these days of material abundance (for many of us), there is little thought given to trashing leftovers, or at least composting dry, stale bits of bread. Yet my friends who've lived in war zones remind me: it's an actual *sin* to waste food. There's a large repertoire of recipes that use up old bread, and these ones are from the real chef in our family, and that's Cheryl. My specialty is winter soups, perhaps because they're so bread-adjacent and snow-loving, as am I. Otherwise, I get to sit back and be the beneficiary of the Italian-Celtic kitchen witchery that goes on every day at our house.

Torta Rustica

Ingredients
Makes 6 servings

Extra-virgin olive oil	3 tablespoons (38 g)
French lentils, dry	½ cup (100 g)
High-quality Roma tomatoes	2 (28-ounce) cans
Salt	1 teaspoon (6 g)
Black pepper	1 teaspoon (2 g)
Dried oregano	1 teaspoon (1 g)
Dried basil	1 teaspoon (1 g)
Leftover bread, cut into ¼- to ½-inch slices	½ loaf
Roughly chopped spinach	3 cups (100 g)
Provolone, goat, and Romano cheese	¼ pound (113 g) each

Torta rustica, translated literally, means "rustic cake." However, rustic *casserole* is a more accurate description for this version, which is made with leftover bread. Semolina-Fennel Bread is best, as it made with semolina flour, slightly denser than other bread. The beauty of this recipe is that it uses up old bread. Also, it is possible to create any number of variations on the theme, using other breads, vegetables, and beans, or even eggs.

Instructions
Preheat the oven to 350°F (176.5°C). Generously coat the bottom of a 13 × 9-inch baking pan with about ⅛ inch of olive oil.

Fill a saucepan with enough water to cover the lentils, plus a couple of inches. Bring the water to a boil, add the lentils, and reduce to a simmer. Cook for about 20 minutes, then test. The lentils should be firm but well cooked. Drain and set aside.

Meanwhile, make the tomato sauce, or "gravy," as Nonna called it. Gravy is more accurate, as tomato sauce is typically thick, while this sauce is almost as thin as juice. Use an immersion (or regular) blender to mix the tomatoes with their juices, olive oil, salt, pepper, oregano, and basil together. The result should be closer to dense tomato juice, thinner than a puree.

Make a layer of old bread slices in the oiled pan. Don't worry about the spaces in between—or, if you prefer, you can square off the slices so they fit more precisely. Next, spread a ladleful of sauce evenly over the bread. Continue layering with the spinach. Add a thin layer of provolone cheese. Next, add the lentils and another ladleful of sauce. Lay more bread slices loosely over the lentils. Add another ladleful of sauce. Goat cheese is next. Layer it about ½ inch thick. Loosely top with more bread and press down a bit. Ladle more sauce over that, then grate a generous layer of Romano cheese on top of everything.

Bake for 30 minutes. The top should be at least golden brown. Let it stand for 10 minutes before serving.

Cheese Lasagne with Bread Noodles

MAKES 6 SERVINGS
For the sauce:

Extra-virgin olive oil	2 tablespoons (25 g)
Mushrooms, stemmed	1½ pounds (680 g)
Small onions, thinly sliced	2
Roasted green or red bell pepper, sliced	1
Large garlic cloves, minced	2
High-quality Roma tomatoes, drained	2 (28-ounce) cans
Dried basil	¼ teaspoon
Dried oregano	¼ teaspoon
Black pepper	¼ teaspoon
Salt	¼ teaspoon

For the lasagne:

Whole-milk ricotta	1 pound (454 g)
Whole-milk mozzarella, shredded	8 ounces (227 g)
Grated Parmesan cheese	½ cup (45 g)
Large eggs	2
Dried basil	¼ teaspoon
Dried oregano	¼ teaspoon
Extra-virgin olive oil	2 tablespoons (25 g)
Leftover bread, cut into ¼-inch or thinner slices	1 loaf

Lasagne is a casserole made by layering extra-wide noodles and cheese with other ingredients. In this recipe, thin slices of bread are substituted for the noodles. Any combination of any variety of bread or ingredients can be used in creating this recipe. However, the ingredients here will be more traditional: thick red sauce with mushrooms, onions, and roasted peppers, along with ricotta, mozzarella, and Parmesan cheeses. The sauce is best when made ahead of time; let cool and then store in the refrigerator until ready to use.

Instructions

Preheat the oven to 350°F (176.5°C).

To make the sauce, heat the oil in a large saucepan over low heat. Add the mushrooms, basil, oregano, and black pepper and sauté for about 5 minutes, until softened. Add the onions and the roasted bell pepper and cook for 5 minutes, then the garlic for 2 minutes more. Add the drained tomatoes, removing as much liquid as possible, and gently crush. Heat for 10 minutes, or until it begins to bubble. Stir in the salt and remove from the heat.

To make the lasagne, combine the ricotta and three-quarters each of the mozzarella and Parmesan together with the eggs in a large bowl. Combine thoroughly, by hand, then season with basil and oregano.

In a 13 × 9-inch baking dish with high sides (at least 3 inches), put a ladle of sauce along with 2 tablespoons olive oil. Begin to layer the ingredients, starting with the slices of bread. Next, add a 1-inch layer of the cheese-egg mixture. Cover with additional slices of bread and another ladle of sauce. Repeat this step three more times. The top layer should be bread, a ladle of sauce, and the remaining mozzarella and Parmesan cheese.

Cover with aluminum foil and bake for 30 minutes, then remove the foil and bake for 20 to 30 minutes more, until golden brown on top.

French Toast

Ingredients
MAKES 4 TO 6 SERVINGS

Colonel Buckwheat Bread, cut into 1-inch slices	1 loaf
Large eggs, room temperature	6
Milk	1½ cups (368 g)
Grated nutmeg	¼ teaspoon
Salt	Pinch
Blueberries (fresh or frozen)	1 quart (680 g)
Maple syrup	½ cup (156 g)

While most of our customers likely think of leftover Challah as the bread to use for French Toast, any loaf can be repurposed for what the French call *pain perdu*, or "lost bread." In fact, it's interesting to include the remains of a variety of loaves. This recipe will use Colonel Buckwheat Bread. A fresh blueberry compote makes it irresistible.

Instructions

Set aside the bread slices or, if you're in a hurry, place them in a 200°F (93°C) oven for about 30 minutes. The goal is to dry out the slices so they will fully absorb the egg mixture.

Beat the eggs in a large bowl, then add the milk, nutmeg, and salt and beat together.

Lay out the bread slices in a 13 × 9-inch baking pan, then pour the egg mixture evenly across the bread. Let it stand to absorb for 15 minutes, then turn each slice over once.

Meanwhile, make the fruit compote. Combine the blueberries and maple syrup in a medium saucepan and heat over low heat for about 5 minutes, stirring frequently.

Heat a large nonstick skillet over medium-low heat or an electric griddle to 350°F (176.5°C). Place the bread slices, not touching, in the skillet and cook until golden, then flip and cook the other side.

Plate your French toast and pour the compote over it.

Rosemary's Stuffing, Baby

Ingredients

MAKES ENOUGH FOR A 15-POUND (6.8 KG) TURKEY

Rosemary Bread, cut into 1-inch cubes	2 loaves
Large eggs	2
Onion, chopped	1
Celery stalk, chopped	1
Small carrot, chopped	1
Garlic clove, minced	1
Salt	¼ teaspoon
Black pepper	¼ teaspoon
Butter	2 tablespoons (28 g)
Extra-virgin olive oil	1 tablespoon (13 g)
Vegetable broth	6 to 8 cups (1½–1¾ L)

There are customers who come into the shop about a week before Thanksgiving with a gleeful (though slightly maniacal) look on their faces. As if they are dying to tell you a secret but don't, they ask for two loaves of Rosemary Bread. The dirty secret is that those loaves headed for the cutting board are not just going to be sliced; they are going to be cubed for stuffing. Undoubtedly destined for a costarring role next to a real turkey.

Instructions

Preheat the oven to 200°F (93°C).

Spread out the bread cubes on parchment-lined baking sheets and bake for 30 minutes. The cubes should be thoroughly dried out. Set aside. Turn the oven up to 300°F (149°C).

In a large bowl, toss together the onion, celery, carrot, garlic, salt, and pepper. Spread the vegetables evenly in a large roasting pan.

Melt the butter with the olive oil over low heat. Pour over the vegetables and mix until evenly covered.

Bake for about 30 minutes, stirring halfway through. This slow roasting should yield lightly caramelized ingredients. Remove from the oven. Turn the oven up to 375°F (190.5°C).

Pour 6 cups vegetable broth into a large bowl. Add the cooked vegetables and dried-out bread and mix well with your hands. Make sure all the cubes have absorbed the broth. Add more broth if not. Allow the mixture to cool to room temperature.

Beat the eggs together in a separate bowl, then add to the cooled mixture and blend it with your hands.

Transfer to a large oiled casserole dish, cover with foil, and bake for 30 minutes. Uncover and bake for an additional 10 minutes, until it is nicely browned on top.

All Aboard the Apprentice-Ship

the starter will smear you
the flour will cake you
the water will splash you
the salt will sting you
the dough will stick you
the scale will tare you
the bench will mark you
the oven will singe you
the steam will scald you
the lame will cut you

and the bread?

The
    bread
       just
          might
            make
              you

whole

Index

Index

Index

About the Author

JONATHAN STEVENS is co-owner of Hungry Ghost Bread in Northampton, MA, nominated six times for the James Beard Awards. His bread has been featured in the *New York Times*, *Boston Globe*, *Saveur*, and *Taste*, among other publications. He has taught baking workshops throughout New England and is ready to sit on your kitchen countertop. He's also a poet, songwriter, and inveterate cyclist. Previous  jobs have included: window washing, rough carpentry, housing advocacy, traditional Inuit medicine research, and merchant marine deckhand in the Gulf of Mexico. His songs can be found on Spotify, and his poems are slipped into bread bags at work. He lives on the edge of Conway State Forest with his partner, Cheryl Maffie, and a sharp bread knife.

Mahesh Ramachandran